Lecture Notes in Mathematics

Edited by A. Dold and B. Eckmann

Subseries: Harvard/MIT
Adviser: G. Sacks

1106

C. T. Chong

Techniques of Admissible Recursion Theory

Springer-Verlag
Berlin Heidelberg New York Tokyo 1984

Author

Chi-Tat Chong
Department of Mathematics, National University of Singapore
Kent Ridge, Singapore 0511, Singapore

AMS Subject Classification (1980): o3D60, O3D25

ISBN 3-540-13902-8 Springer-Verlag Berlin Heidelberg New York Tokyo
ISBN 0-387-13902-8 Springer-Verlag New York Heidelberg Berlin Tokyo

Printed in Germany

Printing and binding: Beltz Offsetdruck, Hemsbach/Bergstr.
2146/3140-543210

FOREWORD

The generalization of recursion theory on the set ω of natural numbers to infinite ordinals (now known as higher recursion theory) has a complicated history. The guiding principle was that basic notions such as computation, finiteness, recursion, recursiveness, relative recursiveness and effectiveness, which lie at the heart of the subject of classical recursion theory, should not be confined to the consideration of ω (and hence countable structures) alone. Indeed in various parts of mathematical logic one had encountered notions and procedures which reminded one of effective computability in the generalized sense. In order to develop a reasonable higher recursion theory, however, it was necessary to have a good intuition of the roles played by objects such as finite sets and r.e. sets in the classical setting. Thus for example in Kreisel [1971] it is aptly pointed out why in any reasonable generalization of recursion theory the notion of 'finiteness' ought to be central. As early as 1938, Kleene had introduced the idea of ordinal notations for natural numbers. This made it possible to discuss certain countably infinite ordinals (the <u>recursive ordinals</u>) in terms of the natural numbers they denote. In the 1950's and early 1960's various recursion-theoretic results on Church-Kleene ω_1 (ω_1^{CK}), the least non-recursive ordinal, were derived, leading to the belief that a recursion theory on this ordinal (later christened 'metarecursion theory' by Kreisel and Sacks) could be developed modelled after classical recursion theory. In particular there should be a correct analog of relative recursiveness which would allow a positive solution to Post's problem for ω_1^{CK}. This was subsequently confirmed in Sacks [1966a], [1966b].

It turns out that Gödel's constructible universe L is an ideal structure in which to develop higher recursion theory. In retrospect one could view the eight fundamental operations introduced by Gödel, in his work on the consistency of the Axiom of Choice and the Generalized Continuum Hypothesis, as recursive functions defined on the universe of sets. Indeed in Takeuti [1960] a study was made on the recursive functions defined on the class of ordinals and it was shown that the cardinals are closed under these functions. Kripke [1964] introduced the notion of admissible ordinals, showing that ω_1^{CK} and all infinite cardinals are instances of these ordinals. He further demonstrated that 'finiteness' in an admissible ordinal could be identified with the property of being an element of L_α (Gödel's L at level α). Closely related to this was Platek's approach [1966] which developed higher recursion theory via definability. He also established the equivalence between basic recursion-theoretic and set-theoretic statements. Recursion theory on admissible ordinals thus evolved into a special case of higher (admissible) recursion theory which could in turn be viewed as a fragment of the Zermelo-Fraenkel set theory. This has since come to be known as the Kripke-Platek set theory. In 1966-67 Sacks introduced the method of priority argument into ordinal recursion theory, solving Post's problem for many admissible ordinals. Finally Jensen's work on the fine structure theory of L ([1972]) provided the necessary tools and insights into admissible recursion theory. Very soon after this, Sacks and Simpson [1972] obtained the complete solution of Post's problem for all admissible ordinals.

This monograph attempts to present an account of the development of the subject of admissible recursion theory that began with the Sacks-Simpson paper. Under the influence of Sacks, higher recursion theory has expanded from admissibility to inadmissibility (Friedman and Sacks [1977]), and into E-recursion (Sacks [1982]). Many of the techniques first introduced to solve problems for admissible ordinals have since been applied to tackle the corresponding ones for inadmissible ordinals. For reason of space, we will devote our attention solely to the admissible case. The emphasis is on introducing the techniques that have been invented to solve problems in the area (hence the title of the book). Each technique is illustrated, as it were, by a result whose proof was the source of the introduction of the particular technique. We give an outline below:

Chapter 1 consists of preliminaries, reviewing some of the basic recursion-theoretic notions and Jensen's fine structure theory. We prove Maass's theorem that (uniformly) there is an α-regular set in every α-r.e. degree. Various characterizations of definable projecta due to Jensen and Simpson are also proved. Chapter 2 discusses forcing over admissible ordinals, and proves Simpson's result on the jump of α-degrees. Chapter 3 introduces the α-finite injury method. We first prove the Sacks-Simpson theorem on the solution of Post's problem, followed by discussing the blocking technique invented by Shore to prove the Splitting Theorem. We introduce the e-state method in Chapter 4, giving Lerman's characterization of α on the existence of maximal sets. This method is studied further in Chapter 5 where we investigate the existence of major subsets (results of Lerman, Leggett-Shore and Chong). In Chapter 6 we study hyperhypersimple sets, and prove existence and non-existence results of these sets (due to Chong and Lerman) on various admissible ordinals. Chapter 7 studies infinite injury method by proving Shore's theorem on the existence of minimal pairs. This study is continued in Chapter 8 where we give a simpler proof of the Density Theorem for all admissible ordinals due originally to Shore. We turn our attention to arbitrary α-degrees in Chapter 9, where we study the existence of minimal degrees and prove the results of Shore and Maass. The notion of the admissible collapse due to Maass is also introduced. In Chapter 10 we turn to set-theoretic methods, where S. Friedman's theorem on the ordering of the degrees above **0'** for many singular cardinals with uncountable cofinality are proved. The book ends with an appendix which presents a proof of Harrington's result that there exists an admissible set for which Post's problem has a negative solution. It is the appropriate place with which to end the treatment since one can say without exaggeration that Post's problem started the whole subject of modern recursion theory.

It should be pointed out that there is an alternative to the approach we have adopted here. Namely, instead of dividing the chapters according to finite injury method, e-state method etc., one could profitably approach it from the point of view of definable projecta. More precisely, various ordinals have been identified to be those where a successful priority argument could be carried out once the set of requirements are indexed within them. For example many finite injury methods can be carried out on requirements of length α^*, the Σ_1 projectum of an admissible ordinal α. The tame Σ_2 projectum turns out to be sufficient to do practically all finite injury arguments. Also Lerman [1974] has shown that the value of the Σ_3 projectum of α is the key to the existence or non-existence of maximal sets. One can argue rather forcefully that this alternate approach brings into sharper focus the contrast between classical and admissible recursion theory, from which the basic features of recursion theory on ω are reflected. Our point of view is somewhat different. The fundamental constructions in α-recursion theory invented thus far consist of those which are discussed in this book. The definable projecta are introduced to ensure that these constructions work. Furthermore recent investigations have concentrated more on specific ordinals where problems have remained unsolved for sometime. As most of these ordinals are constructible cardinals, they are their own definable projecta. This means that the method of coding the set of requirements by a short indexing set fails completely. Nevertheless, some impressive results have surfaced with the use of set-theoretic methods (Friedman [1981]). Some of the most challenging problems in recursion theory today are on the structure of degrees over constructibly singular cardinals (for example the minimal $\aleph_\omega^L$-degree problem). The approach taken here is intended to lay emphasis to this.

We assume that the reader is familiar with classical recursion theory. Lerman [1983], Soare [1984] and Rogers [1967] are recommended for reference material.

Grateful thanks are due to Manuel Lerman, Wolfgang Maass, Richard Shore and Dongping Yang for helpful discussions on the subject matter of this monograph; to Stephen Simpson for providing us with a set of mimeographed notes on introduction to admissible recursion theory; to Sy Friedman for the MIT higher recursion theory lecture notes (1980/1981); to Bob Luborsky for his helpful comments; to Gerald Sacks for lending his kind support to the project and for making a number of helpful suggestions. We also thank the National University of Singapore for awarding a research grant (No. RP84/83), and Tan Bee Hua for her valuable assistance, in the preparation of this volume.

C. T. Chong

1984

Singapore

CONTENTS

CHAPTER 1

INTRODUCTION

We work inside the Zermelo-Fraenkel language of set theory. The following notations will be used throughout. Small Greek letters $\alpha, \beta, \gamma, \ldots$ will always denote (infinite) ordinals. The letter κ is reserved for cardinals, while λ is used only for limit ordinals. Small Roman letters k, m, n, ... will denote natural numbers. An ordinal is identified with its set of predecessors. The cartesian product of two sets x and y is denoted $x \times y$. If f is a function, x a set, then $f|x$ is the function f restricted to x. We write dom(f) for the domain of the function f, while f"x is the image set of x under f.

Our point of departure is the constructible universe of Gödel, denoted L. This is the union of L_γ, as γ ranges over all ordinals. The L_γ's are defined by recursion:

$$L_0 = \emptyset$$

$$L_{\gamma+1} = \{x | x \subseteq L_\gamma \ \& \ x \text{ is first-order definable over } L_\gamma\}$$

$$L_\lambda = \bigcup L_\gamma, \ \gamma < \lambda.$$

Recall the classification of formulas according to Levy [1965]. A formula in the language is Σ_0 (or Π_0) if it contains no unbounded quantifiers (parameters are allowed). These are formulas of the type $(\exists x \varepsilon y)\phi$ or $(\forall x \varepsilon y)\phi$, where ϕ is a quantifier free formula. For $n > 0$, a formula is Σ_n (respectively Π_n) if it is of the form $(\exists x)\phi$ (respectively $(\forall x)\phi$) where ϕ is Π_{n-1} (respectively Σ_{n-1}). A subset x of L_γ is said to be Σ_n definable over L_γ if there is a Σ_n formula $\phi(t)$

(with parameters in L_γ) such that $x = \{u \mid L_\gamma \models \phi(u)\}$. It follows that $L_{\gamma+1}$ is the union of all subsets of L_γ which are Σ_n definable over L_γ for some n.

<u>1.1 DEFINITION</u> Let λ be a limit ordinal. Then λ satisfies the Σ_n replacement axiom if, for every Σ_n formula $\phi(x,y)$, the following holds:

$$L_\lambda \models (\forall u)[(\forall x\varepsilon u)(\exists y)\phi(x,y) \to (\exists z)(\forall x\varepsilon u)(\exists y\varepsilon z)\phi(x,y)].$$

In particular, (1.1) says that if ϕ is a Σ_n function, then for all x in L_λ, the set ϕ"x is an element of L_λ.

<u>1.2 DEFINITION</u> Let α be a limit ordinal. Then α is Σ_n admissible if it satisfies the Σ_n replacement axiom.

A limit ordinal is admissible if it is Σ_1 admissible. The notion of Σ_1 admissibility is of fundamental importance in higher recursion theory. From now on α will always denote an admissible ordinal. The following are two basic facts whose proofs are in Devlin [1974]. They will be used implicitly throughout this book.

(1.3) For every n, there is bijection between α and $\alpha\times\alpha\times\ldots\times\alpha$ (n-fold product).

(1.4) There is a bijection between α and L_α.

By (1.3) one may always regard a (partial) function f: $\alpha\times\alpha\times\ldots\times\alpha \to \alpha$ to be a subset of α. <u>The bijections in (1.3) and (1.4) are both Σ_1 definable over L_α.</u>

<u>1.5 DEFINITION</u> An α-finite set is an element of L_α. A function f: $\alpha \to \alpha$ is (partial) α-recursive if its graph is Σ_1 definable over L_α.

It is not difficult to show that an ordinal α is admissible if and only if

(1.6) for every partial α-recursive function f, for every α-finite set $K \subseteq \text{dom}(f)$, the set $f''K$ is α-finite.

<u>1.7 PROPOSITION</u> There exists a total α-recursive function f of two variables such that for all $\eta < \alpha$, $\{x | f(x,\eta) = 0\}$ is α-finite, and every α-finite set is expressible in this form. Thus there is an effective enumeration of all α-finite sets.

Proof. Let j be the bijection given in (1.4). Let $f(x,\eta)$ be equal to 1 if $x \in j(\eta)$, and let it be 0 otherwise. Since j is Σ_1 definable over L_α, so is f. This f is the desired function.

By (1.4) we will always identify an α-finite set with a bounded set of ordinals below α which is an element of L_α.

<u>1.8 DEFINITION</u> A set $C \subseteq \alpha$ is α-recursively enumerable (abbreviated α-r.e.) if it is the domain of a (partial) α-recursive function.

<u>1.9 DEFINITION</u> There is a partial α-recursive function g of two variables such that for all $e < \alpha$, $\{x | g(x,e) = 0\}$ is α-r.e. and every α-r.e. set is expressible in this form. Hence there is an effective enumeration of all α-r.e. sets.

Proof. There is a Σ_1 definable enumeration of all Σ_1 formulas (with parameters in L_α) of one free variable (cf. Devlin [1974]), Jensen [1972]). Furthermore, the satisfaction relation (over L_α) for Σ_1 sentences is uniformly Σ_1. Let $g(x,e)$ be equal to 1 if $L_\alpha \models \phi_e(x)$, where ϕ_e is the eth Σ_1 formula of one free variable under this enumeration. Let g be undefined otherwise. Then g is the desired function.

From now on, we let K_η and W_e be respectively the ηth α-finite set and the eth α-r.e. set. The two correspondences $\eta \leftrightarrow K_\eta$ and $e \leftrightarrow W_e$ are α-recursive.

The α-finite sets and α-r.e. sets are the basic objects in α-recursion theory. The principal feature of an α-r.e. set is its dynamic growth potential. Every α-r.e. set is built up by a process of enumeration stage by stage. Elements (ordinals) which are enumerated in an α-r.e. set at any stage stay in that set thereafter. At any stage σ of the enumeration, the collection of ordinals enumerated by stage σ form an α-finite set. This can be easily seen in the following way: Suppose that $W = \{x \mid L_\alpha \models (\exists z)\phi(x,z,a)\}$, where ϕ is Σ_1 and $a \in L_\alpha$. Then $W = \bigcup W^\sigma$, where $W^\sigma = \{x \mid L_\alpha \models x < \sigma \;\&\; (\exists z<\sigma)\phi(z,x,a)\}$. Furthermore we have $W^\tau \subseteq W^\sigma$ if $\tau < \sigma$.

1.10 EXAMPLES

(a) ω is admissible. Indeed ω is Σ_n admissible for all n. The ω-finite sets are just the ordinary finite sets of natural numbers, and the ω-r.e. sets are the 'classical' recursively enumerable sets.

(b) ω_1^{CK}, the first non-recursive ordinal (Church-Kleene ω_1), is admissible. The ω_1^{CK}-finite subsets of ω coincide with the hyperarithmetical sets of natural numbers. A set of natural numbers is ω_1^{CK}-r.e. if and only if it is Π^1_1 (Kreisel and Sacks [1965]). This is the first admissible ordinal after ω. There is a one-one total ω_1^{CK}-recursive map p_1 from ω_1^{CK} into ω. The inverse of p_1 is a partial ω_1^{CK}-recursive map and can be extended to a total Σ_2 map from ω onto ω_1^{CK}. Thus ω_1^{CK} is not Σ_2 admissible.

(c) δ^1_2, the least ordinal which is not the order-type of a Δ^1_2 set of natural numbers, is admissible. The δ^1_2-finite sets of ω coincide with the Δ^1_2 sets of integers, and every Σ^1_2 subset of ω is δ^1_2-

r.e. This is the least stable ordinal (i.e. an ordinal α such that L_α is a Σ_1 elementary substructure of L) (Takeuti [1965]).

(d) Every infinite cardinal in a transitive model M of Zermelo-Fraenkel set theory is admissible. This may be seen from the fact that every such infinite cardinal is a stable ordinal in the sense of M, and that every stable ordinal is admissible. Furthermore, every infinite regular cardinal in M is Σ_n admissible for all n. This is not true for singular cardinals. For example, if M is a model of V = L, then the ωth cardinal in M is not Σ_2 admissible. The presence of admissible ordinals which are not Σ_2 admissible makes the subject of admissible recursion theory at once both challenging and intricate. It is the source of many techniques invented in α-recursion theory to settle some fundamental problems in the absence of Σ_2 admissibility.

(e) Various ordinals considered in the study of recursion theory on higher types, such as $\omega_1^{E_1}$, the least ordinal not recursive in E_1 (the type 2 Tugue functional), are admissible (Richter [1968]).

(f) (A non-example) limit ordinals are in general not admissible.

<u>1.11 DEFINITION</u> An α-r.e. set C is α-recursive if and only if C and α - C are α-r.e.

The following proposition is easily verified.

<u>1.12 PROPOSITION</u>

(a) A set C is α-r.e. if and only if it is the range of an α-recursive function.

(b) A set K is α-finite if and only if it is a bounded subset of α and α-recursive.

We now discuss the notion of reducibility. The most important ones considered in classical recursion theory are many-one reducibility, one-one reducibility, truth-table reducibility, and Turing reducibility. We will study a suitable generalization of Turing reducibility in admissible recursion theory.

<u>1.13 DEFINITION</u> Let A, B be subsets of α. B is weakly α-recursive in A (denoted $B \leq_{w\alpha} A$) if there is an α-r.e. set W_e such that for all x,

$$x \in B \quad \leftrightarrow \quad (\exists\nu)(\exists\zeta)[(x,\nu,\zeta,1) \in W_e \ \& \ K_\nu \subseteq A_e \ \& \ K_\zeta \subseteq \alpha - A_e].$$

$$x \notin B \quad \leftrightarrow \quad (\exists\nu)(\exists\zeta)[(x,\nu,\zeta,0) \in W_e \ \& \ K_\nu \subseteq A_e \ \& \ K_\zeta \subseteq \alpha - A_e].$$

If a set B is weakly α-recursive in A via W_e (as in 1.13), we write $\{e\}^A = B$, and say that B is weakly α-recursive in A via the reduction procedure e. We say that B is α-r.e. in A if there is a function $f \leq_{w\alpha} A$ such that B is the domain of f. As in classical recursion theory, 'domain' may be replaced by 'range'.

When $\alpha = \omega$, weak α-recursiveness coincides with (or, perhaps, is also known as) Turing reducibility. In this case the relation is denoted $\leq_T$ and is reflexive and transitive. This results in the partitioning of subsets of ω into various equivalence classes called Turing degrees. For arbitrary α such a partition need not exist. Indeed Driscoll [1968] showed that for $\alpha = \omega_1^{CK}$ the relation $\leq_{w\alpha}$ is not transitive even on α-r.e. sets. Shore [1975b] has characterized the admissible ordinals α for which $\leq_{w\alpha}$ is not transitive on the α-r.e. sets to be precisely those with more than one incomplete, non-hyperregular α-r.e. degree (see below for definition). Intuitively one sees where the problem lies: Suppoose $C = \{d\}^B$ and $B = \{e\}^A$. To compute C from A element-wise, it would seem natural to combine d and e together to form a new reduction procedure. However, by 1.13 it takes α-finitely much information about B to decide whether an element x belongs to C. This information consists of two parts: one which is a subset of B and one which is a subset of α − B. Let these be H and K respectively. Then to decide from A whether x is in C, one would have to locate, using A, the α-finite sets H and K. Now while each element of H and K may be found from C in α-finite time, using only α-finitely much information about A, their totality need not. The point is that the union of an α-finite set need not be α-finite. To overcome this difficulty, we introduce the following stronger notion of reducibility.

1.14 DEFINITION B is α-recursive in A if there is an α-r.e. set W_e such that for all γ,

$$K_\gamma \subseteq B \quad \leftrightarrow \quad (\exists\nu)(\exists\zeta)[(\gamma,\nu,\zeta,1) \in W_e \;\&\; K_\nu \subseteq A \;\&\; K_\zeta \subseteq \alpha - A].$$

$$K_\gamma \subseteq \alpha - B \quad \leftrightarrow \quad (\exists\nu)(\exists\zeta)[(\gamma,\nu,\zeta,0) \in W_e \;\&\; K_\nu \subseteq A \;\&\; K_\zeta \subseteq \alpha - A].$$

This relation is denoted $\leq_\alpha$ and we say that B is α-recursive in A via the reduction procedure e. We sometimes also write $\{e\}^A = B$, with the understanding that computations are performed on α-finite sets rather than just on individual elements.

There is an analog of Church's thesis in admissible recursion theory. Stated simply, this says that a set B can be 'effectively' computed from a set A, in the sense that there is an algorithm to decide from A whether a given α-finite set is contained in B (or in α − B), if and only if there is an e such that B is α-recursive in A via the reduction procedure e. We do not attempt to define here what is meant by the word 'effective'. Suffice it to say that, at the cost of involving one with messy formal details, every statement in this book which mentions 'effectively computable from' meets the criterion set by Definition 1.14.

The following is easy to verify.

<u>1.15 PROPOSITION</u> The relation $\leq_\alpha$ is transitive. Furthermore, if $B \leq_\alpha A$ then $B \leq_{w\alpha} A$.

<u>1.16 DEFINITION</u> If $B \leq_\alpha A$ and $A \leq_\alpha B$, then A and B have the same α-degree.

We write $A \equiv_\alpha B$ to mean A and B have the same α-degree. Small Roman letters **a**, **b**, **c**...are used to denote α-degrees. The α-degree of α-recursive sets is denoted **0**. As in classical recursion theory, the α-degrees form an upper semilattice: Given **a** and **b**, we let $\mathbf{a} \cup \mathbf{b}$ be the α-degree of the set

$$A \oplus B = \{2x \mid x \in A\} \cup \{2x+1 \mid x \in B\},$$

where the α-degree of A is **a** and the α-degree of B is **b**.

There is a third notion of reducibility, that of α-calculability (Sacks [1967]). Roughly this says that B is α-calculable from A if it is possible to decide the elements of B from the deduction tree that starts with a finite set in the equation calculus for α together with the diagram of A. The main point to note here is that each deduction, in contrast with those in the other two notions of reducibility, need not be α-finite, even though it may mention only ordinals less than α. In this monograph we will concentrate our attention mainly on α-recursive reducibility, the most elaborately studied notion of reducibility in admissible recursion theory.

<u>1.17 DEFINITION</u> $A \subseteq \alpha$ is α-regular if $A \cap \gamma$ is α-finite for every $\gamma < \alpha$.

This is an important notion in generalized recursion theory. An α-recursive set is clearly α-regular. Assuming $V = L$, every subset of an infinite cardinal κ is κ-regular. There are however numerous examples of admissible ordinals α with α-r.e. sets which are not α-regular. The existence of non-α-regular α-r.e. sets shows that in general even bounded α-r.e. sets may require α-many steps to enumerate.

1.18 EXAMPLES

(a) Let $\alpha = \omega_1^{CK}$. Let $\mathcal{O}$ be Kleene's complete Π_1^1 set. Then by 1.10 (b) $\mathcal{O}$ is an α-r.e. subset of ω. This set is not α-finite, hence not α-regular.

(b) In general, if α is not a constructible cardinal, then there is a subset of $|\alpha|^L$ (the L-cardinality of α) which is not α-finite, hence not α-regular. A natural example would be a subset of $|\alpha|^L$ of order-type α. This set is not α-finite since otherwise there would be an α-recursive map from $|\alpha|^L$ onto α, violating admissibility (1.6).

There is however a representation theorem due to Sacks (Sacks [1966a]) which states that every α-degree of an α-r.e. set contains an α-regular α-r.e. set (see also Simpson [1971]). Maass [1978b] refined the proof to show that such a representation could be found uniformly. Before we prove this, let us first introduce the notion of the Σ_1 projectum.

1.19 DEFINITION The Σ_1 projectum of α is the least ordinal γ for which there is a $\Sigma_1(L_\alpha)$ definable subset of γ which is not α-finite.

1.20 THEOREM The following are equivalent:

(a) γ is the Σ_1 projectum of α

(b) γ is the least ordinal δ for which there is a partial α-recursive function taking a subset of δ onto α

(c) γ is the least ordinal δ for which there is a total α-recursive injection of α into δ.

Proof. Suppose that γ is the Σ_1 projectum of α. Let W_e be an α-r.e. subset of γ which is not α-finite. For each $\sigma < \alpha$, let $g(\sigma)$ be the least x (in the order of the given enumeration for W_e) such that x is enumerated in W_e at some stage $\tau > \sigma$ and x is not equal to $g(\sigma')$ for all $\sigma' < \sigma$. It can be seen that g is $\Sigma_1(L_\alpha)$, and hence α-recursive. Observe that g is a total function. For suppose that $g(\sigma)$ is not defined. Then this means that every element of W_e has been enumerated by stage σ, implying that W_e is α-finite, a contradiction. Thus g is total and so (a) implies (c).

Now suppose that (c) is true. Let f be a total α-recursive function mapping α one-one into γ. Then the inverse of f is a partial α-recursive function taking a subset of γ onto α. Hence (c) implies (b).

Finally if h is a partial α-recursive function mapping a subset of γ onto α, then the domain of h is α-r.e. Furthermore it is not α-finite by (1.6). This shows that (b) implies (a).

The Σ_1 projectum of α is denoted α^*. It is also known simply as the projectum.

1.21 REMARKS

(a) The projectum of ω_1^{CK} is ω. By Theorem 1.20 the projectum of an infinite cardinal is itself. The latter is essentially the crucial fact used in Gödel's proof of the consistency of the GCH.

(b) The proof of (a) $\rightarrow$ (c) in Theorem 1.20 shows that every non-α-finite α-r.e. set is the range of a one-one total α-recursive function. We will use this fact implicitly in the sequel.

We now prove Maass's theorem. From now on α-regular sets will be abbreviated regular sets if the context is clear.

1.22 THEOREM There is an α-recursive function r such that for all $e < \alpha$, $W_e \equiv_\alpha W_{r(e)}$ and $W_{r(e)}$ is regular.

Proof. Let $P: \alpha \to \alpha^*$ be an α-recursive injection given by Theorem 1.20. Let W_e be α-r.e. and let

$$A = \{2\gamma \mid K_\gamma \cap W_e \neq \emptyset\} \cup \{2\gamma+1 \mid K_\gamma \subseteq W_e\}.$$

Then A is α-r.e. and has the same α-degree as W_e. It is sufficient to consider the case when W_e is not α-finite. Let f be a one-one total α-recursive function which enumerates A. Let

$$B = \{(\gamma,x) \mid (\exists y>x)[f(y) < \gamma \ \& \ P(f(y)) < P(f(x))\}.$$

$B = W_{r(e)}$ is the 'deficiency set' of A and it is easily seen that the function r is α-recursive. We claim that B and W_e have the same α-degree and B is regular.

$W_e \leq_\alpha B$. Let K be an α-finite set. Then $K = K_\gamma$ for some γ. To decide whether $K \subseteq W_e$ it is sufficient to see whether $2\gamma+1 \in A$. Also to see whether $K_\gamma \subseteq \alpha - W_e$ it is sufficient to see whether $2\gamma \in A$. We will use B to make such decisions. Let $z = 2\gamma$ or $2\gamma+1$, depending on the case we have in mind. Firstly note that there is an unbounded set of ordinals x such that $(z,x) \in B$. To see this, let an ordinal t be fixed. Pick the $x > t$ such that $P(f(x))$ is minimal. Then by the definition of B we see that $(z,x) \notin B$. Since α^* is the projectum of α and $P(z) < \alpha^*$, the set $\{P(f(u)) \mid P(f(u)) < P(z)\}$ is an α-r.e. set bounded below α^*, and hence α-finite. It follows that the set $\{u \mid P(f(u)) < P(z)\}$ is α-finite, hence bounded below α. Thus there is an x such that $(z,x) \notin B$ and $P(f(x)) > P(z)$. Then z is not in A unless

$z \in f''x+1$. Hence $W_e \leq_\alpha B$.

$B \leq_\alpha W_e$. By examining the definition of B, we have $(\gamma,y) \notin B$ if and only if for every $z < \gamma$ such that $P(z) < P(f(y))$, either $z \notin A$ or $z \in f''y+1$. It is then possible to use A to decide if a given pair (γ,y) belongs to B. Now given an α-finite set K, one can go through every pair (γ,y) in K to decide if it belongs to B. This procedure ends at a stage before α. We have $B \leq_\alpha W_e$.

B is regular. Let $\beta < \alpha$. We show that

$$B \cap \{(\gamma,z) \mid \gamma < \beta \ \& \ z < \beta\}$$

is α-finite. Define the following two sequences:

γ_1 = the least ordinal $\gamma < \beta$ for which there is a $y \geq \beta$ with $f(y) < \gamma$

y_1 = the $y \geq \beta$ such that $f(y) < \gamma_1$ and $P(f(y))$ is minimal

γ_{n+1} = the least $\gamma > \gamma_n$ such that $\gamma < \beta$ and there is a $y > \beta$ for which $f(y) < \gamma$ and $P(f(y)) < P(f(y_n))$

y_{n+1} = the y satisfying the definition of γ_{n+1} such that $P(f(y))$ is minimal.

If γ_1 does not exist, then clearly

$$B \cap \{(\gamma,x) \mid \gamma < \beta > \ \& \ x < \beta\}$$

is empty. We may therefore assume that γ_1 and hence y_1 exist. Now we have $P(f(y_1)) > P(f(y_2)) > \ldots$, a decreasing sequence of ordinals, so that there is an n where γ_n is not defined, which then implies that y_n is not defined. Let n_0 be the least such n. Let $\gamma_{n_0} = \beta$. This means that for all $y > y_{n_0}$, if $P(f(y)) < P(f(y_{n_0}))$, then $f(y) > \beta$. Now let

(γ,x) be given such that $\gamma_j < x < \gamma_{j+1}$. If (γ,x) is in B, then either there is a $y < \beta$ such that $y > x$, $f(y) < \gamma$, and $P(f(y)) < P(f(x))$, or else there is a $y > \beta$ such that $f(y) < \gamma$ with $P(f(y)) < P(f(x))$. In the latter case the y where $P(f(y))$ is minimal is one of the y_i's. This gives an effective algorithm for computing elements of B in $\beta\times\beta$. Hence B is regular. The proof of Theorem 1.22 is complete.

Using Skolem hull argument, Maass [1978b] also showed that there is a parameter-free Σ_1 function r such that for all admissible ordinals α and for all $e < \alpha$, $W_e \equiv_\alpha W_{r(e)}$ and $W_{r(e)}$ is regular.

We now turn to the fine structure theory of L. A good reference for this is Devlin [1974a]. This theory, first developed by Jensen, has found wide and important applications in higher recursion theory (cf. S. Friedman [1984] for an up-to-date survey). We will not give a detailed account of this theory, but will only discuss those aspects of the theory which have proved most relevant to α-recursion theoretic considerations.

<u>1.23 DEFINITION</u> The Σ_n projectum of α, denoted $\sigma np(\alpha)$, is the least ordinal γ for which there is a $\Sigma_n(L_\alpha)$ definable subset of γ which is not α-finite.

Thus $\sigma 0p(\alpha) = \alpha$ by the admissibility of α. When $n = 1$, we have the notion of the projectum defined previously. We will prove an analog of Theorem 1.20 for all $n < \omega$. This requires a much more intricate argument, involving the idea of Σ_n Skolem hulls together with the Σ_n Uniformization Theorem.

<u>1.24 DEFINITION</u> Let $\rho < \alpha$. The $\Sigma_n(L_\alpha)$ cofinality of ρ, denoted $\sigma ncf_\alpha(\rho)$, is the least $\gamma < \rho$ for which there is a $\Sigma_n(L_\alpha)$ function mapping γ cofinally into ρ.

When $\rho = \alpha$, we will drop the subscript α in $\sigma ncf_\alpha(\alpha)$. The ordinal $\sigma ncf(\alpha)$ measures the extent to which α is not Σ_n admissible. It is clear that $\sigma ncf(\alpha) = \alpha$ if and only if α is Σ_n admissible.

<u>1.25 DEFINITION</u> An α-cardinal is an ordinal $\gamma < \alpha$ which is a cardinal in the sense of L_α, i.e. there is no α-finite bijection between γ and a smaller ordinal.

Localizing the argument that constructible cardinals are admissible, one can show that every α-cardinal is admissible.

The following is easy to verify.

<u>1.26 PEOPOSITION</u> For all n, if $\sigma np(\alpha) < \alpha$, then it is an α-cardinal. The same is true of $\sigma ncf_\alpha(\rho)$ for all $\rho < \alpha$.

Proof. If $\sigma np(\alpha) < \alpha$ is not a cardinal in the sense of L_α, then there is an α-finite bijection $f\colon \sigma np(\alpha) \to \gamma$, for some $\gamma < \sigma np(\alpha)$. Let $A \subseteq \sigma np(\alpha)$ be $\Sigma_n(L_\alpha)$ and not α-finite. Then f"A is a $\Sigma_n(L_\alpha)$ subset of γ and not α-finite, a contradiction.

Now suppose that $\sigma ncf_\alpha(\rho)$ is not an α-cardinal. Let $f\colon \sigma ncf_\alpha(\rho) \to \gamma$ be an α-finite bijection, where $\gamma < \sigma ncf_\alpha(\rho)$. Let $g\colon \sigma ncf_\alpha(\rho) \to \rho$ be $\Sigma_n(L_\alpha)$ and cofinal in ρ. Then $f \circ g^{-1}\colon \gamma \to \rho$ cofinally and is $\Sigma_n(L_\alpha)$, contradicting the choice of $\sigma ncf_\alpha(\rho)$.

We next state Jensen's Σ_n Uniformization Theorem. This is a result which illustrates on the one hand how a level by level analysis of the Σ_n hierarchy is carried out, and on the other hand how recursion theoretic applications of fine structure theory are made possible. The theorem says roughly that for every $\Sigma_n(L_\alpha)$ relation there is a $\Sigma_n(L_\alpha)$ choice function (without increasing the complexity of the definition!). For n = 1 this amounts to saying that an α-r.e. relation has a (partial) α-recursive choice function, a fact which is not difficult to prove. Surprisingly the proof for arbitrary n as originally given in Jensen [1972] (or Devlin [1974a]) is very complicated. We present a simplification due to S. Friedman (cf. Simpson [1978]). It should be noted that even this simplified proof utilizes some sophisticated ideas.

1.27 THE Σ_n UNIFORMIZATION THEOREM Let $n \geq 1$. For each $\Sigma_n(L_\alpha)$ relation $\phi(x,y)$, there is a $\Sigma_n(L_\alpha)$ function f such that for all $x < \alpha$, if there is a y such that $L_\alpha \models \phi(x,y)$, then $L_\alpha \models \phi(x,f(x))$.

We first give an application of Theorem 1.27. This result will be used extensively in the monograph.

1.28 THEOREM Assume the Σ_n Uniformization Theorem. Then $\sigma np(\alpha)$ is the least $\gamma \leq \alpha$ for which there exists a one-one $\Sigma_n(L_\alpha)$ function $p_n: \alpha \to \gamma$. By taking the inverse of p_n, we also have $\sigma np(\alpha)$ to be the least γ for which there is a one-one partial $\Sigma_n(L_\alpha)$ function from a subset of γ onto α. We call p_n a $\Sigma_n(L_\alpha)$ projection.

Proof. Suppose $\gamma < \alpha$ such that $p: \alpha \to \gamma$ is a one-one $\Sigma_n(L_\alpha)$ function. Let $A = \{p(\eta) | \eta \notin K_\eta\}$. Then A is a $\Sigma_n(L_\alpha)$ subset of γ. Furthermore, A is not α-finite, since if $A = K_\eta$, then $p(\eta) \in A$ if and only if $p(\eta) \notin K_\eta = A$, a contradiction. Thus $\sigma np(\alpha) \leq \gamma$.

Thus to prove the theorem it is sufficient to show that there is a $\Sigma_n(L_\alpha)$ projection of α into $\sigma np(\alpha)$. We may assume that $\sigma np(\alpha) < \alpha$. First of all, generalizing the proof of Proposition 1.9, it can be shown that there exists a partial $\Sigma_n(L_\alpha)$ function of two variables $g(x,e)$ such that for all $e < \alpha$, $\{x | g(x,e) = 0\}$ is $\Sigma_n(L_\alpha)$, and every $\Sigma_n(L_\alpha)$ subset of α is of this form. Hence there is a $\Pi_{n-1}(L_\alpha)$ formula $\phi(x,z,e)$ such that for all e and for all x, $g(x,e) = 0$ if and only if $(\exists z)\phi(x,z,e)$. By Theorem 1.27, there is a $\Sigma_n(L_\alpha)$ function f such that for all e and for all x, $g(x,e) = 0$ if and only if

$$\phi(x,z) \quad \leftrightarrow \quad \text{for all } y <_n z,\ L_\alpha \models \psi(x,y,z).$$

Proof. For n = 0 the lemma is easily seen to be true since p_0 in this case is the identity function, so that checking whether $\phi(x,z)$ holds is an α-recursive procedure. Thus assume $n > 0$. By induction suppose that the lemma is true for $0,\ldots, n-1$. We have

$$\begin{aligned}\phi(x,z) \quad &\leftrightarrow \quad (\forall y<_n z)\psi(x,y,z)\\ &\leftrightarrow \quad (\forall y<_n z)(\exists u)\psi'(x,y,z,u),\end{aligned}$$

where ψ' is $\Pi_{n-1}(L_\alpha)$. We consider two cases:

Case 1. $\sigma ncf_\alpha(\sigma(n-1)p(\alpha)) < \sigma np(\alpha)$. Thus there is a $\Sigma_n(L_\alpha)$ function f mapping a $\kappa < \sigma np(\alpha)$ cofinally into $\sigma(n-1)p(\alpha)$. Then $\phi(x,z)$ holds over L_α if and only if for all $y <_n z$, there is a $w < \kappa$ such that $\psi'(x,y,z,u)$ holds over L_α for some u such that $p_{n-1}(u) < f(w)$. Now the relation

$$\begin{aligned}R(p_n(y),w) \quad \leftrightarrow \quad & p_n(y) < p_n(z)\ \&\ w < \kappa\ \&\\ & (\exists u)[p_{n-1}(u) < f(w)\ \&\ \psi'(x,y,z,u)]\end{aligned}$$

is easily seen to be a $\Sigma_n(L_\alpha)$ subset of $p_n(z)\times\kappa$. By Proposition 1.26 $\sigma np(\alpha)$ is either equal to α or is an α-cardinal, therefore admissible. Hence the pairing function (which is $\sigma np(\alpha)$-recursive) assures that R is bounded below $\sigma np(\alpha)$. By 1.23 we have R to be α-finite. We may then uniformize R by an α-finite function g. This means that $\phi(x,z)$ holds if and only if there is an α-finite function g such that for all $y <_n z$ there is a u such that $p_{n-1}(u) < f\circ g(p_{n-1}(y))$ such that $\psi'(x,y,z,u)$ holds, i.e.

$$\begin{aligned}\phi(x,z) \quad \leftrightarrow \quad & (\exists g)(\forall y)[y <_n z \rightarrow (\exists u)(p_{n-1}(u) <\\ & f\circ g(p_{n-1}(y))\ \&\ \psi'(x,y,z,u))]\end{aligned}$$

As by induction $(\exists u)(P_{n-1}(u) < f\circ g(p_{n-1}(y))\ \&\ \psi'(x,y,z,u))$ is $\Pi_n(L_\alpha)$

$\phi(x,f(x),e)$. Now let $A \subseteq \sigma np(\alpha)$ be $\Sigma_n(L_\alpha)$ and non-α-finite. Let X_0 be α-finite which contains the initial segment of ordinals less than $\sigma np(\alpha)$, and the parameters used to define both A and the Σ_n uniformizing function f (say a and b respectively). Form the $\Sigma_n(L_\alpha)$ Skolem hull of X_0 in L_α. Let this be X. Then every element of X is Σ_n definable from some parameters in X_0. There is in fact a $\Sigma_n(L_\alpha)$ (Skolem) function h with parameters a and b such that $h"X_0 = X$. It follows that there is a $\Sigma_n(L_\alpha)$ function h' such that $(h')"\sigma np(\alpha) = X$. Let $\pi: X \to L_\beta$ be a bijection, where β = the least ordinal not in X (π is the 'condensation map'). By 1.4 we may in fact regard π to be the map taking X to β. By the fact that X is the $\Sigma_n(L_\alpha)$ Skolem hull of X_0, one can verify that the map π is actually $\Sigma_n(L_\alpha)$. We now have $h^* = \pi \circ h$: $\sigma np(\alpha) \to \alpha$, a $\Sigma_n(L_\alpha)$ map from a subset of $\sigma np(\alpha)$ onto β, with parameters $\pi(a)$ and $\pi(b)$. Now observe that for all $x < \sigma np(\alpha)$, $h^*(x) = x$, by our choice of X_0. It follows that $(h^*)"A = A$, and so A is $\Sigma_n(L_\beta)$ definable using parameter $\pi(a)$. If $\beta < \alpha$, we would get A to be α-finite, contary to assumption. Hence $\beta = \alpha$. Thus h^* is a $\Sigma_n(L_\alpha)$ function from a subset of $\sigma np(\alpha)$ onto α. Define $\psi(y,x)$ if and only if $h^*(x) = y$. Let p_n be a Σ_n uniformizing function for ψ. Then p_n is a Σ_n projection. Taking inverse we have a one-one injection from a subset of $\sigma np(\alpha)$ onto α.

We now turn to the proof of Theorem 1.27. To fix notations, let $p_n: \alpha \to \sigma np(\alpha)$ be a $\Sigma_n(L_\alpha)$ projection, as provided by Theorem 1.28. Given x, y less than α, write $x <_n y$ if $p_n(x) < p_n(y)$. Clearly the relation $x <_n y$ is $\Sigma_n(L_\alpha)$. For $n = 0$, p_n may be chosen to be the identity function. We first prove a lemma.

<u>1.29 LEMMA</u> Suppose that the $\Sigma_k(L_\alpha)$ Uniformization Theorem holds for all k, $1 \leq k \leq n$. Let $\psi(x,y,z)$ be $\Sigma_n(L_\alpha)$. Then the relation $\phi(x,z)$ is $\Sigma_{n+1}(L_\alpha)$, where

(note that $\psi'(x,y,z,u)$ is $\Pi_{n-1}(L_\alpha)$), we see that $\phi(x,z)$ is $\Sigma_{n+1}(L_\alpha)$.

Case 2. $\sigma ncf_\alpha(\sigma(n-1)p(\alpha)) \geq \sigma np(\alpha)$. Since $\phi(x,z)$ holds if and only if $(\forall y<_n z)(\exists u)\psi'(x,y,z,u)$ holds, we know by the Σ_n Uniformization Theorem that there is a $\Sigma_n(L_\alpha)$ function g such that

$$\phi(x,z) \quad \leftrightarrow \quad (\forall y<_n z)\psi'(x,y,z,g(y)).$$

We now have a $\Sigma_n(L_\alpha)$ map h taking $p_n(y)$ to $p_{n-1}(g(y))$, a map with domain some $\Sigma_n(L_\alpha)$ subset of $p_n(z) < \sigma np(\alpha)$ (hence α-finite by 1.23) and range contained in $\sigma(n-1)p(\alpha)$. As the domain of h is α-finite, we might as well consider it to be a proper initial segment of $\sigma np(\alpha)$. But then by the assumption on $\sigma ncf_\alpha(\sigma(n-1)p(\alpha))$, the image of h is bounded below $\sigma(n-1)p(\alpha)$ by some $p_{n-1}(s)$. Hence we have

$$\phi(x,z) \quad \leftrightarrow \quad (\exists s)(\forall y<_n z)(\exists u<_{n-1} s)\psi'(x,y,z,u).$$

Again this is easily seen to be $\Sigma_{n+1}(L_\alpha)$. This proves the lemma.

<u>Proof of the Uniformization Theorem</u>

We proceed by induction on n.

Let n = 1. Then as noted above $\phi(x,y)$ is an α-r.e. relation. It is then possible to α-recursively enumerate pairs (x,y) such that $\phi(x,y)$ holds. Let f(x) = y if and only if (x,y) occurs in this enumeration and (x,y) is the least such pair in the enumeration for the given x. Then f is (partial) α-recursive and so the theorem is true for n = 1.

Now assume that the theorem is true for all $k \leq n$. We show that it is true for n+1. Let $\phi(x,y)$ be $\Sigma_{n+1}(L_\alpha)$ be defined by

$$\phi(x,y) \quad \leftrightarrow \quad (\exists z)\psi(x,y,z),$$

where $\psi(x,y,z)$ is $\Pi_n(L_\alpha)$. Let

$$\psi'(x,y,z) \leftrightarrow \psi(x,y,z) \ \& \ (\forall y',z')[(y',z') <_n (y,z) \rightarrow \sim\psi(x,y,z)].$$

Thus $\psi'(x,y,z)$ holds if and only if (y,z) is the p_n-least pair which makes $\psi(x,y,z)$ true. By Lemma 1.29 ψ' is $\Sigma_{n+1}(L_\alpha)$. Then the function f defined by

$$f(x) = y \leftrightarrow (\exists z)\psi'(x,y,z)$$

is $\Sigma_{n+1}(L_\alpha)$ and is the choice function for $\phi(x,y)$. This proves Theorem 1.27.

At this point we turn our attention to the topic of relativization. For $A \subseteq \alpha$, consider the structure $\langle L_\alpha, A\rangle$. One may expand the language of set theory to include a predicate for A. It is then possible to define sets which are Σ_n definable in the structure $\langle L_\alpha, A\rangle$. These are called $\Sigma_n(L_\alpha,A)$ definable sets. They may be given very natural recursion theoretic interpretation.

<u>1.30 PROPOSITION</u> Let $A \subseteq \alpha$ be a regular set. Then $B \subseteq \alpha$ is $\Delta_1(L_\alpha,A)$ if and only if $B \leq_{w\alpha} A$.

Proof. We give a sketch. Suppose that B is $\Delta_1(L_\alpha,A)$. Then there exist $\Delta_0(L_\alpha,A)$ formulas ϕ and ψ such that for all $x < \alpha$,

$$x \in B \leftrightarrow (\exists y)\phi(y,x), \text{ and}$$

$$x \notin B \leftrightarrow (\exists y)\psi(y,x).$$

The formulas ϕ and ψ, being $\Delta_0(L_\alpha,A)$, have bounded quantifiers. These may, for example, be of the form $(\forall z<y)\theta(z,x)$ or $(\exists z<y)\theta(z,x)$, where θ is quantifier free. Since A is regular, expressions of the form

$$(\forall z<y)(z \in A \rightarrow \ldots)$$

are therefore decided by only α-finitely much information about A. It follows that there is an effective procedure to decide from A whether an ordinal x is in B. Hence $B \leq_{w\alpha} A$.

Now suppose that $B \leq_{w\alpha} A$. Then there is an e such that for all $x < \alpha$,

$$x \in B \leftrightarrow (\exists\eta)(\exists\nu)[(x,\eta,\nu,1) \in W_e \ \& \ K_\eta \subseteq A \ \& \ K_\nu \subseteq \alpha - A], \text{ and}$$

$$x \notin B \leftrightarrow (\exists\eta)(\exists\nu)[(x,\eta,\nu,0) \in W_e \ \& \ K_\eta \subseteq A \ \& \ K_\nu \subseteq \alpha - A].$$

Since α-r.e. sets are by definition $\Sigma_1(L_\alpha)$, the above expressions may be represented by $\Sigma_1(L_\alpha,A)$ formulas (note that $K_\eta \subseteq A$ and $K_\nu \subseteq \alpha - A$ are $\Delta_0(L_\alpha,A)$.

<u>1.31 COROLLARY</u> Let A be regular. A total function f is $\Sigma_1(L_\alpha,A)$ if and only if $f \leq_{w\alpha} A$.

Partial functions which are weakly α-recursive in A are also called partial α-A-recursive functions. B is α-r.e. in A if B is the range (or domain) of a function weakly α-recursive in A. This is equivalent to saying that there is an e such that for all $x < \alpha$,

$$x \in B \leftrightarrow (\exists\eta)(\exists\nu)[(x,\eta,\nu,1) \in W_e \ \& \ K_\eta \subseteq A \ \& \ K_\nu \subseteq \alpha - A].$$

<u>1.32 DEFINITION</u> $A \subseteq \alpha$ is α-hyperregular (hyperregular for short) if $\langle L_\alpha,A\rangle$ is an admissible structure, i.e. for every $\Sigma_1(L_\alpha,A)$ formula $\phi(x,y)$, for every set u in $\langle L_\alpha,A\rangle$,

$$\langle L_\alpha,A\rangle \models (\forall x\varepsilon u)(\exists y)\phi(x,y) \rightarrow (\exists v)(\forall x\varepsilon u)(\exists y\varepsilon v)\phi(x,y).$$

The next result is easy to verify.

<u>1.33 PROPOSITION</u> Let A be regular and hyperregular. Then B is $\Delta_1(L_\alpha, A)$ if and only if $B \leq_\alpha A$.

Proposition 1.33 indicates the strength of an hyperregular set in terms of the theory of ordinal recursion. Combining it with Proposition 1.30 we see that $\leq_{w\alpha}$ and $\leq_\alpha$ coincide for regular and hyperregular sets. This happens for instance when $\alpha = \omega$. There are, however, many instances of sets which are not hyperregular. If $\alpha^* < \alpha$ and $p_1 \colon \alpha \to \alpha^*$ is an α-recursive projection, then $p_1"\alpha$ is an α-r.e., non-hyperregular set. This set is also not regular. On the other hand, the set of non-cardinals in $\aleph_\omega^L$ (the ωth constructible cardinal) is a regular α-r.e. set which is not hyperregular.

In contrast with regularity (at least for α-r.e. sets), the property of hyperregularity is degree-theoretically invariant. This means that if $A \equiv_\alpha B$ are α-r.e. sets, then A is hyperregular if and only if B is hyperregular. Non-hyperregularity is the source of many fundamental problems in admissible recursion theory. For example, the problem of the existence of a minimal degree for $\aleph_\omega^L$ is not known, mainly because the 'complete' α-r.e. set $\{(x,e) \mid x \in W_e\}$ is not hyperregular (we discuss this further in Chapter 9). Let $A \subseteq \alpha$ be regular. To measure the extent to which a structure $\langle L_\alpha, A\rangle$ is not admissible, one introduces the notion of <u>$\Sigma_1(L_\alpha, A)$ cofinality</u>. This is the least ordinal γ for which there is a function weakly α-recursive in A mapping γ cofinally into α. This ordinal, denoted $\sigma 1cf(\alpha, A)$, is also known as <u>the recursive cofinality of A</u> (Stillwell [1969]). This property is again degree-theoretically invariant. Thus one has $\sigma 1cf(\alpha, A) = \alpha$ if and only if A is hyperregular.

Closely related to the notion of $\Sigma_1(L_\alpha, A)$ cofinality is that of the $\Sigma_1(L_\alpha, A)$ projectum, denoted $\sigma 1p(\alpha, A)$ (also $\alpha^*(A)$ in analogy with the notation α^*). This is the least γ for which there is a $\Sigma_1(L_\alpha, A)$

subset which is not α-A-finite (i.e. not an element of $\langle L_\alpha, A\rangle$). For A regular and $\sigma 1cf(\alpha, A) \geq \sigma 1p(\alpha, A)$, this is also the least γ for which there is a total α-A-recursive function from α into γ, or the least γ for which there is a partial α-A-recursive function from a subset of γ onto α. These facts will be used quite often in later chapters. Their verification is left to the reader.

1.34 DEFINITION Let $n \geq 0$. A $\Sigma_n(L_\alpha)$ master code for α is a set $A \subseteq \sigma np(\alpha)$ such that for all $m \geq 1$, $B \subseteq \sigma np(\alpha)$ is $\Sigma_{m+n}(L_\alpha)$ if and only if it is $\Sigma_m(L_\alpha, A)$.

We will prove that $\Sigma_n(L_\alpha)$ master codes exist for all $n \geq 0$, and we will choose a canonical sequence of such sets to be denoted A_n. For $\alpha = \omega$, the $\Sigma_n(L_\alpha)$ master codes should remind the reader of the Σ_n complete sets $\emptyset^{(n)}$. The master codes are defined in a similar way, but with a somewhat more set-theoretic flavour. The following proposition follows easily from Definition 1.23.

1.35 PROPOSITION Let $A \subseteq \sigma np(\alpha)$ be $\Sigma_n(L_\alpha)$. Then A is $\sigma np(\alpha)$-regular.

1.36 PROPOSITION Let A_n be a $\Sigma_n(L_\alpha)$ master code. Then $\sigma(n+1)p(\alpha)$ = the least ordinal γ for which there is a $\Sigma_1(L_\alpha, A_n)$ subset of γ which is not α-finite, i.e. $\sigma(n+1)p(\alpha) = \sigma 1p(\alpha, A_n)$.

Proof. By 1.23, $\sigma(n+1)p(\alpha) \leq \sigma np(\alpha)$, and so by 1.34 for any subset of $\sigma(n+1)p(\alpha)$, it is $\Sigma_{n+1}(L_\alpha)$ if and only if it is $\Sigma_1(L_\alpha, A_n)$. This holds in particular for the ones which are not α-finite.

1.37 THEOREM For each $n \geq 0$, there is a $\Sigma_n(L_\alpha)$ master code.

Proof. We prove this by induction on n.

For $n = 0$, $\sigma 0p(\alpha) = \alpha$ by admissibility of α, and one may simply take A_0 to be α.

Let $\rho = \sigma n p(\alpha)$ and $\rho' = \sigma(n+1)p(\alpha)$. Let A_m be a $\Sigma_m(L_\alpha)$ master code for all $m < n$. Let $\{\phi_i(x)\}_{i<\omega}$ be a $\Sigma_1(L_\alpha, A_n)$ enumeration of all $\Sigma_1(L_\alpha, A_n)$ foymulas. Set

$$A_{n+1} = \{(i,x) \mid x < \rho' \ \& \ \langle L_\rho, A_n \rangle \models \phi_i(x)\}.$$

It can be verified that the satisfaction relation

$$\langle L_\alpha, A_n \rangle \models \phi_i(x)$$

is $\Sigma_1(L_\rho, A_n)$ (uniformly over i) and so A_{n+1} is $\Sigma_1(L_\rho, A_n)$, hence $\Sigma_{n+1}(L_\alpha)$ and ρ'-regular. We claim that A_{n+1} is a $\Sigma_{n+1}(L_\alpha)$ master code.

Let C be a subset of ρ' and Σ_1 over $\langle L_{\rho'}, A_{n+1} \rangle$. By Corollary 1.31 and Proposition 1.35, C is ρ'-r.e. in A_{n+1}. Hence there is a d such that for all x,

$$(1.38) \quad x \in C \leftrightarrow (\exists \eta)(\exists \nu)[(x,\eta,\nu,1) \in W_d \ \& \ K_\eta \subseteq A_{n+1} \ \& \ K_\nu \subseteq \rho' - A_{n+1}].$$

But A_{n+1} is $\Sigma_1(L_\rho, A_n)$, and hence ρ-r.e. (again by 1.31 and 1.35). So there is an e such that for all y,

$$(1.39) \quad y \in A_{n+1} \leftrightarrow (\exists \eta)(\exists \nu)[(y,\eta,\nu,1) \in W_e \ \& \ K_\eta \subseteq A_n \ \& \ K_\nu \subseteq \rho - A_n].$$

We consider two cases:

Case 1. $\langle L_\rho, A_n \rangle$ is admissible.

Then for given K_η, one has by admissibility

$$(1.40) \quad K_\eta \subseteq A_n \leftrightarrow (\exists \tau)(\forall y \in K_\eta)(\exists \eta, \nu < \tau)[(y,\eta,\nu,1) \in W_e \ \& \ K_\eta \subseteq A_n \ \& \ K_\nu \subseteq \rho - A_n].$$

Combining (1.38), (1.39) and (1.40) we have a $\Sigma_2(L_\rho,A_n)$ definition for C. Hence every $\Sigma_1(L_{\rho'},A_{n+1})$ set is $\Sigma_2(L_\rho,A_n)$.

Case 2. $\langle L_\rho,A_n\rangle$ is not admissible.

Let $\kappa = \sigma lcf(\rho,A_n)$. Let f be weakly ρ-A_n-recursive mapping $\sigma lcf(\rho,A_n)$ cofinally into ρ (thus f is $\Sigma_1(L_\rho,A_n)$. By assumption $\kappa < \rho$. Then $K_\zeta \subset A_{n+1}$ if and only if

$$(1.41)\quad (\forall z\varepsilon K_\zeta)(\exists\gamma<\kappa)(\exists\eta,\nu<f(\gamma))[(z,\eta,\nu,1) \; \varepsilon \; W_e \;\& \; K_\eta \subseteq A_n \;\&\; K_\nu \subseteq \rho - A_n].$$

Since the domain of f is all of κ (hence α-finite), one may verify as in Proposition 1.30 that f is weakly ρ-recursive in A_n, so that f is $\Delta_1(L_\rho,A_n)$. The expression $(\exists\eta,\nu<f(\gamma))[\ldots\;\;]$ in 1.41 is then $\Delta_1(L_\rho,A_n)$. Define $T(y,\gamma)$ to hold if and only if $y \; \varepsilon \; K_\zeta$, $\gamma < \kappa$, and there exist η, $\nu < f(\gamma)$ such that $(y,\eta,\nu,1) \; \varepsilon \; W_e$, with $K_\eta \subseteq A_n$ and $K_\nu \subseteq \rho - A_n$. Then T is $\Delta_1(L_\rho,A_n)$ and bounded below ρ' (notice that K_ζ is by our choice ρ'-finite and $\kappa < \rho'$). As A_n is a $\Sigma_n(L_\alpha)$ master code, T is in fact $\Sigma_{n+1}(L_\alpha)$, hence must be α-finite. Thus there is an α-finite function $g\colon K_\zeta \to \kappa$ such that $T(y,\gamma)$ holds for $y \; \varepsilon \; K_\zeta$ and some $\gamma < \kappa$ if and only if $T(y,g(y))$ holds. By Proposition 1.26 ρ' is an α-cardinal (if it is less than α; the other case is trivial), and so by stability every α-finite subset bounded below ρ' is in fact ρ'-finite. Thus $K_\zeta \subseteq A_{n+1}$ if and only if

$$(1.42)\quad (\exists g)(\forall y\varepsilon K_\zeta)(\exists\eta,\nu<f(g(y)))[(y,\eta,\nu,1) \; \varepsilon \; W_e \;\& \; K_\eta \subseteq A_n \;\&\; K_\nu \subseteq \rho - A_n].$$

This is clearly $\Sigma_1(L_\rho,A_n)$. Combining (1.38), (1.39) and (1.42) one concludes that C is $\Sigma_2(L_\rho,A_n)$. It follows that $\Sigma_1(L_{\rho'},A_{n+1})$ is a subset of $\Sigma_2(L_\rho,A_n)$.

By induction, one can show that in general for $r > 0$,

$\Sigma_r(L_{\rho'}, A_{n+1})$ is contained in $\Sigma_{r+1}(L_\rho, A_n)$. We leave this to the reader.

To prove the other direction, let C be $\Sigma_{r+1}(L_\rho, A_n)$ such that C is a subset of ρ'. To be specific, suppose that r is even. Then

$$(1.43) \quad x \in C \quad \leftrightarrow \quad (\exists y_1)(\forall y_2)\ldots(\forall y_r)\phi(x,y_1,\ldots,y_r),$$

where ϕ is $\Sigma_0(L_\rho, A_n)$. Now by Theorem 1.28 there is a $\Sigma_{n+1}(L_\alpha)$ injection p_{n+1} of α into ρ'. The restriction q of this map to ρ is then a $\Sigma_1(L_\rho, A_n)$ injection of ρ into ρ'. Let $\phi'(x,y_1',\ldots,y_r')$ be defined by

$$(1.44) \quad x < \rho' \ \& \ y_1',\ldots,y_r' \in \text{range}(q) \ \& \\ \phi(x,q^{-1}(y_1'),\ldots,q^{-1}(y_r')).$$

By the definition of A_{n+1}, B = range(q) is actually Δ_1 in $\langle L_{\rho'}, A_{n+1}\rangle$, and therefore so is ϕ'. Then $x \in C$ if and only if

$$(1.45) \quad (\exists y_1' \in B)(\forall y_2' \in B)\ldots(\forall y_r' \in B)\phi(x,y_1',\ldots,y_r'),$$

showing that C is $\Sigma_r(L_{\rho'}, A_{n+1})$. We conclude that A_{n+1} is a $\Sigma_{n+1}(L_\alpha)$ master code.

A notion closely related to that of the $\Sigma_n(L_\alpha)$ projectum is the $\Delta_n(L_\alpha)$ projectum.

<u>1.46 DEFINITION</u> The $\Delta_n(L_\alpha)$ projectum is the least ordinal γ for which there is a $\Delta_n(L_\alpha)$ subset of α which is not α-finite. This ordinal is denoted $\delta np(\alpha)$.

As every $\Delta_n(L_\alpha)$ set is $\Sigma_n(L_\alpha)$, we have $\sigma np(\alpha) \leq \delta np(\alpha) \leq \delta(n-1)p(\alpha)$. The relative sizes of these ordinals can be described accurately. This follows as a consequence of the next theorem.

<u>1.47 THEOREM</u> The $\Delta_n(L_\alpha)$ projectum is the least ordinal γ for which there is a $\Delta_n(L_\alpha)$ surjection of γ onto α.

Proof. Let δ be the least ordinal γ for which there exists a $\Delta_n(L_\alpha)$ surjection of γ onto α. The proof that $\delta np(\alpha) \leq \delta$ is similar to that given in the characterization of $\sigma np(\alpha)$ (Theorem 1.28). We show here that the other inequality also holds.

<u>Claim</u> $\sigma ncf_\alpha(\sigma(n-1)p(\alpha)) \leq \delta np(\alpha)$.

Let $B \subset \delta np(\alpha) = \rho$ be $\Delta_n(L_\alpha)$ and not α-finite. Let A_{n-1} be a $\Sigma_{n-1}(L_\alpha)$ master code (Theorem 1.37). Then B is a $\Delta_n(L_\alpha)$ subset of ρ and therefore $\Sigma_1(L_\rho, A_{n-1})$. Then there exist $\Delta_0(L_\rho, A_{n-1})$ formulas ϕ and ψ such that B is defined by

$$x \in B \quad \leftrightarrow \quad (\exists y)\phi(x,y)$$

$$x \notin B \quad \leftrightarrow \quad (\exists y)\psi(x,y).$$

Let $f\colon \delta np(\alpha) \to \rho$ be defined so that

$$f(x) = y \quad \leftrightarrow \quad (\phi(x,y) \text{ or } \psi(x,y)) \;\&\; (\forall z)[z < y \to (\sim\phi(x,y) \;\&\; \sim\psi(x,y))].$$

By Theorem 1.37 the function f is $\Sigma_n(L_\alpha)$. Now the range of f is not bounded below ρ since otherwise we have a $\gamma < \rho$ such that for all $x < \delta np(\alpha)$, there is a $y < f(\gamma)$ for which $\phi(x,y)$ or $\psi(x,y)$ holds, showing that B is in fact an element of $\langle L_\rho, A_{n-1}\rangle$. This would in turn imply that B is α-finite, a contradiction. Thus the range of f is cofinal in ρ.

Now by Theorem 1.28, there is a $\Sigma_n(L_\alpha)$ projection $p_n\colon \alpha \to \rho'$, where $\rho' = \sigma np(\alpha)$. The restriction of this projection to ρ then gives a $\Sigma_n(L_\alpha)$ map taking ρ one-one into ρ'. This map, which we denote by g, is then $\Sigma_1(L_\rho, A_{n-1})$. Then we have $g(x) = y$ if and only if $(\exists z)\theta(x,y,z)$, where θ is $\Delta_0(L_\rho, A_{n-1})$. Let f be a $\Sigma_n(L_\alpha)$ function taking $\delta np(\alpha)$ cofinally into ρ, as provided by the claim. For $u, v < \delta np(\alpha)$, define

$$h(u,v) = \begin{cases} x & \text{if } (\exists z<f(v))\theta(x,u,z) \\ 0 & \text{otherwise} \end{cases}$$

The function h is $\Sigma_1(L_\rho, A_{n-1})$, and therefore $\Sigma_n(L_\alpha)$. It is furthermore total on $\delta np(\alpha)\times\delta np(\alpha)$. By (1.3) we now have a $\Sigma_n(L_\alpha)$ function h' from $\delta np(\alpha)$ onto ρ. Taking the inverse of p_{n-1} (as given by Theorem 1.28), we may make it into a $\Sigma_n(L_\alpha)$ function q from ρ onto α. Now the composition $q\circ h'$ is the desired function.

Again using Theorem 1.27 one may show that $\delta np(\alpha)$ is also the least ordinal γ for which there is a $\Delta_n(L_\alpha)$ function from α onto γ.

1.48 COROLLARY $\delta np(\alpha) = \max\{\sigma np(\alpha), \sigma ncf_\alpha(\sigma(n-1)p(\alpha)\}$.

Proof. As the proof of Theorem 1.47 shows, if γ is an ordinal such that $\sigma np(\alpha) \leq \gamma \leq \sigma(n-1)p(\alpha)$, and there is a $\Sigma_n(L_\alpha)$ cofinal map of γ into $\sigma(n-1)p(\alpha)$, then there is a $\Delta_n(L_\alpha)$ function mapping γ onto α. This implies by Theorem 1.47 that in this case $\delta np(\alpha)$ is in fact equal to $\sigma ncf_\alpha(\sigma(n-1)p(\alpha))$. Now suppose that $\sigma ncf_\alpha(\sigma(n-1)p(\alpha)) < \sigma np(\alpha)$. Then again by the proof of Theorem 1.47 there is a $\Delta_n(L_\alpha)$ function from $\delta np(\alpha)$ onto α. Then $\delta np(\alpha) \leq \sigma np(\alpha)$. But the inequality cannot be strict, by the definition of $\sigma np(\alpha)$. It follows that in this case $\sigma np(\alpha) = \delta np(\alpha)$.

Theorem 1.28 turns out to be very useful for recursion-theoretic purposes. Theorems 1.37, 1.47 and Corollary 1.48 may be generalized to those levels L_β of the constructible hierarchy which are 'rudimentarily closed' (thus without the assumption of admissibility). In this case all the results continue to hold with essentially the same proof (removing the admissibility assumption necessitates a split into cases). These are some of the key features of the Jensen hierarhy J_β and its related hierarchy S_β ([1974a], Jensen [1972]). We will have occassion to make use of the S-hierarchy in Chapter 10 where set-theoretic methods in α-recursion theory are discussed. It is convenient there to simply think of S_β as L_β for β a limit ordinal.

Lerman [1974] introduced the notion of an $S_n(L_\alpha)$ projectum in his solution of the maximal set problem. To define this, let us say that for a function $h(\sigma,x)$ of two variables, $\lim_\sigma h(\sigma,x)$ exists and is equal to $h(x)$ if $h(\sigma,x) = h(x)$ for all sufficiently large σ.

<u>1.49 DEFINITION</u> A function f is $S_n(L_\alpha)$ if there is an α-recursive function $f'(\sigma_1,\dots,\sigma_{n-1},x)$ such that

$$f(x) = \lim_{\sigma_1} \lim_{\sigma_2} \dots \lim_{\sigma_{n-1}} f'(\sigma_1,\dots,\sigma_{n-1},x).$$

It is easy to verify that a function is $\Sigma_1(L_\alpha)$ if and only if it is $S_1(L_\alpha)$.

<u>1.50 THEOREM</u> A function is $\Sigma_2(L_\alpha)$ if and only if it is $S_2(L_\alpha)$.

Proof. Clearly every $S_2(L_\alpha)$ function is $\Sigma_2(L_\alpha)$. We prove that the converse holds. Let f be $\Sigma_2(L_\alpha)$ so that

$$f(x) = y \quad \leftrightarrow \quad (\exists w)(\forall z)\phi(x,y,w,z),$$

where ϕ is $\Delta_1(L_\alpha)$, hence α-recursive. Define

$$g(\sigma,x,y) = \text{the least } w < \sigma \text{ such that } (\forall z<\sigma)\phi(x,y,w,z),$$

where 'least' is taken with the understanding that if such a w does not exist, then $g(\sigma,x,y) = \sigma$. Set

$$f'(\sigma,x) = \text{the least } y < \sigma \text{ such that } (\forall z<\sigma)\phi(x,y,g(x,y,z),z).$$

We claim that $\lim_\sigma f'(\sigma,x) = f(x)$. To see this, suppose $f(x) = y$. Let w_0 be the least w such that $(\forall z)\phi(x,y,w,z)$. Now for all $y' < y$, there is by admissibility a $\sigma(y')$ such that

$$g(\sigma(y'),x,y') \neq w \text{ for all } w < w_0.$$

Another application of admissibility gives a σ_0 such that $g(\sigma_0,x,y') \neq w'$ for all $w' < w_0$. Then at stage σ_0, one has $f'(\sigma_0,x) = y$, and indeed by the definition of g, we see that $f'(\sigma,x) = y$ for all $\sigma > \sigma_0$. Hence f is $S_2(L_\alpha)$.

Clearly every $S_n(L_\alpha)$ function is $\Sigma_m(L_\alpha)$ for some m. The converse is however false. Simpson [1978] gave an example of an admissible ordinal α where a $\Sigma_3(L_\alpha)$ function fails to be $S_n(L_\alpha)$ for all $n < \omega$.

1.51 DEFINITION A subset of α is $S_n(L_\alpha)$ if its characteristic function is $S_n(L_\alpha)$.

1.52 DEFINITION The $S_n(L_\alpha)$ projectum of α is the least ordinal γ for which there exists an $S_n(L_\alpha)$ subset of γ which is not α-finite.

The following result is a consequence of Theorem 1.50.

1.53 COROLLARY $S_2(L_\alpha)$ projectum $= \delta 2p(\alpha)$.

Proof. By Theorem 1.50 every $S_2(L_\alpha)$ function is $\Sigma_2(L_\alpha)$. This means that its graph is $\Delta_2(L_\alpha)$. The conclusion now follows from the definition of $\delta 2p(\alpha)$.

<u>1.54 COROLLARY</u> $S_2(L_\alpha)$ projectum is the least ordinal γ for which there is an $S_2(L_\alpha)$ function mapping γ onto α.

Proof. This follows from Theorems 1.47 and 1.53.

Corollary 1.54 points out a possible Jensen type characterization for the $S_n(L_\alpha)$ projectum. This is in fact true, as proved by Simpson (cf. Simpson [Notes]).

<u>1.55 THEOREM</u> The $S_n(L_\alpha)$ projectum is the least ordinal γ for which there is an $S_n(L_\alpha)$ function from γ onto α.

Proof. The case for n = 1 is easy to verify. Corollary 1.54 proves the case for n = 2. We will only prove the theorem for n = 3, the one needed for the proof of Lerman's theorem in Chapter 4.

<u>Claim 1</u> If $\sigma 2cf(\alpha) = \delta 2p(\alpha)$, then a function is $\Sigma_3(L_\alpha)$ if and only if it is $S_3(L_\alpha)$.

Let f be $S_3(L_\alpha)$. Then there is an α-recursive function f' such that

$$f(x) = y \quad \leftrightarrow \quad \lim_\sigma \lim_\tau f'(\sigma,\tau,x) = y.$$

Unravelling the definition, we get

$$f(x) = y \quad \leftrightarrow$$
$$(\exists\sigma_0)(\forall\sigma>\sigma_0)(\exists\sigma')(\forall\tau>\sigma')f'(\sigma,\tau,x) = y.$$

The expression $(\forall\tau>\sigma)f'(\sigma,\tau,x) = y$ is $\Pi_1(L_\alpha)$. Then $P(\sigma,\sigma')$ defined by

$$P(\sigma,\sigma') \quad \leftrightarrow \quad (\forall\tau>\sigma')f'(\sigma,\tau,x) = y$$

is $\Sigma_2(L_\alpha)$, and so by the Uniformization Theorem 1.27 there is a $\Sigma_2(L_\alpha)$ function g such that $P(\sigma,g(\sigma))$ holds whenever $P(\sigma,\sigma')$ holds for some

σ'. We then have

$$f(x) = y \quad \leftrightarrow$$
$$(\exists \sigma_0)(\forall \sigma>\sigma_0)(\forall \tau)[\tau > g(\sigma) \rightarrow f'(\sigma,\tau,x) = y].$$

As the graph of g is $\Delta_2(L_\alpha)$, we see that f is $\Sigma_3(L_\alpha)$.

Now let f be $\Sigma_3(L_\alpha)$. Then there is a $\Pi_2(L_\alpha)$ formula ϕ such that

$$f(x) = y \quad \leftrightarrow \quad (\exists u)\phi(x,y,u).$$

Since $\sigma 2cf(\alpha) = \delta 2p(\alpha) = S_2(L_\alpha)$ projectum by Corollary 1.53, there is an $S_2(L_\alpha)$ surjection $h: \delta 2p(\alpha) \rightarrow \alpha$. Set

$$j(x) = \text{the least } (y,u) < \sigma 2cf(\alpha) \text{ such that } \phi(x,h(y),h(u)) \text{ holds.}$$

Then $f(x) = h(j(x)_1)$, where $j(x)_1$ denotes the first component of $j(x)$. By Theorem 1.50 the function h is $S_2(L_\alpha)$. Also the function taking a pair (y,u) to y is α-recursive, hence $S_n(L_\alpha)$ for all n. Now it is not difficult to see that the $S_n(L_\alpha)$ functions are closed under composition. Thus to show that f is $S_3(L_\alpha)$ it is sufficient to show that the function j is $S_3(L_\alpha)$. Now $j(x) = (y,u)$ is $\Pi_2(L_\alpha)$, so we may define the following approximation:

$$j'(\sigma,x) = \text{the least } z < \sigma 2cf(\alpha) \text{ such that } (\forall v)(\exists w)\psi(x,v,w,z);$$

$$j''(\sigma,\tau,x) = \text{the least } z < \sigma 2cf(\alpha) \text{ such that } (\forall v<\sigma)(\exists w<\tau)\psi(x,v,w,z).$$

Here $\psi(x,v,w,z)$ is an α-recursive predicate. Now observe that for σ fixed, $\lim_\tau j''(\sigma,\tau,x) = j'(\sigma,x)$ by admissibility. Furthermore, if $u < j(x)$ then by the definition of $j(x)$ we know that

$(\exists v)(\forall w){\sim}\psi(x,v,w,u)$

The relation

$$R(u,\sigma) \quad \leftrightarrow \quad u < x \ \& \ (\exists v{<}\sigma)(\forall w){\sim}\psi(x,v,w,u)$$

is $\Sigma_2(L_\alpha)$ and so by the Uniformization Theorem (1.27) there is a $\Sigma_2(L_\alpha)$ function r such that $R(u,r(u))$ holds for all $u < j(x)$. Since $j(x) < \sigma 2cf(\alpha)$, we have a uniform bound σ_0 such that

$$(\forall u{<}j(x)(\exists v{<}\sigma_0)(\forall w){\sim}\psi(x,v,w,u).$$

We then have $j'(\sigma,x) = \min\{\sigma,j(x)\}$. It follows that $j(x)$ is equal to $\lim_\sigma j'(\sigma,x)$. This proves Claim 1.

<u>Claim 2</u> If $\sigma 2cf(\alpha) < \delta 2p(\alpha)$, then $S_{n+1}(L_\alpha)$ projectum is equal to $\sigma 2p(\alpha) = \delta 2p(\alpha)$.

Choose a $\Sigma_2(L_\alpha)$ cofinal function $f\colon \sigma 2cf(\alpha) \to \alpha$. We show that every $S_{n+1}(L_\alpha)$ set bounded below $\sigma 2p(\alpha)$ is α-finite.

Let X be an $S_{n+1}(L_\alpha)$ set bounded below $\beta < \sigma 2p(\alpha)$. Let g' be α-recursive taking values in 2 so that

$$X(x) = \lim_{\sigma_1} \lim_{\sigma_2} \ldots \lim_{\sigma_n} g'(\sigma_1,\ldots,\sigma_n,x),$$

where we have identified X with its characteristic function, so that $X(x) = 1$ if $x \in X$, and $X(x) = 0$ otherwise. Define $g^*(\tau_1,\ldots,\tau_n,x) = g'(f(\tau_1),\ldots,f(\tau_n),x)$. Then the graph g^* is $\Sigma_2(L_\alpha)$ and bounded below $\sigma 2p(\alpha)$, hence α-finite. We then have

$$X(x) = \lim_{\tau_1} \ldots \lim_{\tau_n} g^*(\tau_1,\ldots,\tau_n,x),$$

showing that X is α-finite.

Now let X be $\Sigma_2(L_\alpha)$ and bounded below $\delta 2p(\alpha)$. Then there is an α-recursive predicate θ such that

$$x \in X \quad \leftrightarrow \quad (\exists y)(\forall z)\theta(x,y,z).$$

Let

$$R(x,\sigma) \quad \leftrightarrow \quad (\exists y<f(\sigma))(\forall z)\theta(x,y,z).$$

It is not difficult to see that R is in fact $\Delta_2(L_\alpha)$, hence being a bounded subset of $\delta 2p(\alpha)$, it is α-finite. It follows that X is α-finite as well.

We may now prove the theorem for n = 3. As before (cf. Theorem 1.28) the difficult direction is in the proof that there is an $S_3(L_\alpha)$ function from the $S_3(L_\alpha)$ projectum onto α. We consider two cases:

Case 1 $\sigma 2cf(\alpha) = \sigma 2p(\alpha)$.

By Claim 1, $S_3(L_\alpha)$ projectum $= \sigma 3p(\alpha)$, and every $\Sigma_3(L_\alpha)$ function is $S_3(L_\alpha)$. By Theorem 1.28 we have the conclusion.

Case 2 $\sigma 2cf(\alpha) < \sigma 2p(\alpha)$.

By Claim 2 one has $S_3(L_\alpha)$ projectum $= \delta 2p(\alpha)$, and by Corollary 1.53 this is in turn equal to $S_2(L_\alpha)$ projectum. By Theorem 1.28 we have the result.

We end this chapter with one more notion of projectum first introduced in Lerman [1972]. This is an ordinal which plays a fundamental role in finite injury priority arguments.

1.56 DEFINITION A function $f: \gamma \to \alpha$ is tame $\Sigma_2(L_\alpha)$ if there exists an α-recursive function f' such that for all $\beta < \gamma$, there is a σ_0 such that

$$f(x) = y \leftrightarrow (\forall\sigma > \sigma_0) f'(\sigma, x) = y,$$

for all $x < \beta$.

Clearly every tame $\Sigma_2(L_\alpha)$ function is $\Sigma_2(L_\alpha)$. Observe, however, that the strength of the definition lies in the existence of a uniform bound σ_0 for the convergence of an arbitrary proper initial segment of γ. It is not difficult to see that every $\Sigma_2(L_\alpha)$ function defined on $\sigma 2cf(\alpha)$ is tame $\Sigma_2(L_\alpha)$.

<u>1.57 DEFINITION</u> A set is tame $\Sigma_2(L_\alpha)$ if its characteristic function is tame $\Sigma_2(L_\alpha)$.

<u>1.58 DEFINITION</u> The tame $\Sigma_2(L_\alpha)$ projectum, denoted $t\sigma 2p(\alpha)$, is the least ordinal γ for which there is a tame $\Sigma_2(L_\alpha)$ subset which is not α-finite.

The following characterization of $t\sigma 2p(\alpha)$ is due to Simpson (unpublished).

<u>1.59 THEOREM</u> $t\sigma 2p(\alpha)$ is the least ordinal γ for which there is a tame $\Sigma_2(L_\alpha)$ surjection of γ onto α.

Proof. We will prove only the more difficult direction, namely that there is a tame $\Sigma_2(L_\alpha)$ surjection from $t\sigma 2p(\alpha)$ onto α. Let A be a tame $\Sigma_2(L_\alpha)$ subset of $t\sigma 2p(\alpha)$ which is not α-finite. We identify A with its characteristic function. Let A* be an α-recursive approximation of A satisfying the tameness condition. Then for all $z < t\sigma 2p(\alpha)$, there is a σ_0 such that

$$A(x) = y \leftrightarrow (\forall\sigma > \sigma_0)(A^*(\sigma, x) = y),$$

for all $x < z$. Let $g(z)$ be the least such σ_0. Then it is easy to see that g is $\Sigma_2(L_\alpha)$. By Theorem 1.28 and the remark following it, let h be $\Sigma_2(L_\alpha)$ mapping a subset of $t\sigma 2p(\alpha)$ onto α. By Theorem 1.50 there is an α-recursive function h' such that

$$h(x) = y \quad \leftrightarrow \quad (\exists\sigma)(\forall\tau>\sigma)(h'(\tau,x) = y).$$

Set $f((x,\nu)) = h'(g(\nu),x)$. Then

$$\begin{aligned} f(u) = v \quad &\leftrightarrow \quad (\exists x)(\exists\nu)[u = (x,\nu) \ \& \ v = h'(g(\nu),x)] \\ &\leftrightarrow \quad (\exists t)(\exists\sigma)(\forall\tau>\sigma)[g'(\tau,\nu) = t \ \& \\ &\qquad v = h'(g'(\tau,\nu),x)], \end{aligned}$$

where g' is an α-recursive approximation of g. The reader may verify that f is tame $\Sigma_2(L_\alpha)$.

The next theorem is also due to Simpson.

<u>1.60 THEOREM</u> $t\sigma 2p(\alpha) = \sigma 2p(\alpha)$ or $\sigma 2p(\alpha).\sigma 2cf(\alpha)$.

Proof. The proof as given in Chong and Lerman [1976] is incomplete (it actually only proves the following claim).

<u>Claim</u> $t\sigma 2p(\alpha) = \delta 2p(\alpha)$ or $\delta 2p(\alpha).\sigma 2cf(\alpha)$.

By Theorem 1.47, let f be $\Sigma_2(L_\alpha)$ mapping $\delta 2p(\alpha)$ onto α. Let f' be its α-recursive approximation. Let $g\colon \sigma 2cf(\alpha) \to \alpha$ be a $\Sigma_2(L_\alpha)$ cofinality function. Then the map $h(\delta 2p(\alpha).\nu+\delta) = f'(g(\nu),\delta)$ gives a tame $\Sigma_2(L_\alpha)$ surjection from $\delta 2p(\alpha).\sigma 2cf(\alpha)$ onto α.

Now suppose that $\delta 2p(\alpha) < t\sigma 2p(\alpha)$. Then $t\sigma 2p(\alpha) = \delta 2p(\alpha).\nu+\delta$ for some $\nu < \sigma 2cf(\alpha)$ and some $\delta < \delta 2p(\alpha)$ (if $\nu = \sigma 2cf(\alpha)$ then $\delta = 0$ by the above). Suppose that $\nu < \sigma 2cf(\alpha)$. If $\delta > 0$, choose a stage σ_0 such that $f'(\sigma,x) = f(x)$ for all $x < \delta 2p(\alpha).\nu$ and all $\sigma > \sigma_0$. Such a stage σ_0 exists by the tameness of f. It follows that $f''\delta 2p(\alpha).\nu$ is an α-finite set bounded by some ordinal γ. Hence there is a tame $\Sigma_2(L_\alpha)$ function mapping $\delta < \delta 2p(\alpha)$ onto a final segment of α

which must have order type α. But clearly using this one may construct a tame $\Sigma_2(L_\alpha)$ function from δ onto α, contradicting the fact that $\delta < \delta 2p(\alpha)$. Thus $\delta = 0$.

Next observe that ν is a limit ordinal. For suppose that $\nu = \gamma+1$. Then $f''\delta 2p(\alpha).\gamma$ is α-finite and so bounded below α. It follows that there is a tame $\Sigma_2(L_\alpha)$ function mapping $\delta 2p(\alpha)$ onto a final segment of α. This map can be modified to become a tame $\Sigma_2(L_\alpha)$ map from $\delta 2p(\alpha)$ onto α, contadicting our hypothesis that $\delta 2p(\alpha) < t\sigma 2p(\alpha)$. Hence ν is a limit ordinal.

Now since $\nu < \sigma 2cf(\alpha)$, one may define a function u: $\nu \to \alpha$ such that $u(\zeta) = \sup_{\rho<\delta 2p(\alpha)} f(\delta 2p(\alpha).\zeta+\rho)$. The function u is $\Sigma_2(L_\alpha)$ and by the tameness of f, $u(\zeta)$ is defined for every $\zeta < \nu$. Moreover u is cofinal in α because f is. This again is a contradiction since $\nu < \sigma 2cf(\alpha)$. We have proved the claim.

We may now prove the theorem. If $\sigma 2p(\alpha) = \delta 2p(\alpha)$, then the theorem follows immediately from the claim. Suppose that $\sigma 2p(\alpha) < \delta 2p(\alpha)$. Then clearly $t\sigma 2p(\alpha) \geq \delta 2p(\alpha) > \sigma 2p(\alpha)$. By Claim 2 of Theorem 1.55, $\sigma 2cf(\alpha) \geq \delta 2p(\alpha)$. Suppose that $\delta 2p(\alpha) = t\sigma 2p(\alpha)$. Then the tame $\Sigma_2(L_\alpha)$ function from $\delta 2p(\alpha)$ onto α already provides a $\Sigma_2(L_\alpha)$ cofinal function from $\delta 2p(\alpha)$ into α, so that indeed $\sigma 2cf(\alpha) = \delta 2p(\alpha)$. Then $t\sigma 2p(\alpha) = \sigma 2cf(\alpha) = \sigma 2p(\alpha).\sigma 2cf(\alpha)$. On the other hand, suppose that $\delta 2p(\alpha) < \sigma 2cf(\alpha)$. Then $t\sigma 2p(\alpha) \neq \delta 2p(\alpha)$ (else we would get a $\Sigma_2(L_\alpha)$ cofinal function from $\delta 2p(\alpha)$ into α) and so by the claim it must be $\delta 2p(\alpha).\sigma 2cf(\alpha) = \sigma 2cf(\alpha) = \sigma 2p(\alpha).\sigma 2cf(\alpha)$. This proves the theorem.

In Chapter 9 we will have occassion to consider admissible structures $\langle L_\delta, T\rangle$, for $T \subseteq \delta$ a regular set, and their tame Σ_2 projectum denoted $t\sigma 2p(\delta,T)$. The relativized version of Theorem 1.60 continues to hold.

CHAPTER 2

THE JUMP OPERATOR AND 1-GENERIC SETS

We devote this chapter to the study of (Cohen) forcing over admissible ordinals. The problem which we will consider for this purpose is that of a suitable generalization of the Friedberg Completeness Theorem. In the classical setting, this theorem states that every Turing degree above **0'** is the jump of some Turing degree. To study the corresponding problem in the setting of admissible ordinals α, one needs for a start a reasonable definition for the jump of an arbitrary subset of α. This in itself does not come by easily, due mainly to the presence of non-regular sets and non-hyperregular sets. There are several possible inequivalent definitions for the jump operator, all of which coincide for classical recursion theory. The difficulty lies in choosing one which is the most natural. Fortunately all of the definitions which appear natural turn out to be equivalent for sets which are regular and hyperregular.

<u>2.1 DEFINITION</u> Let $B \subseteq \alpha$. The eth set α-r.e. in B is the set

$$W_e^B = \{x \mid (\exists\nu)(\exists\eta)[(\nu,\eta,x,1) \in W_e \ \&\ K_\nu \subseteq B \ \&\ K_\eta \subseteq \alpha - B]\}.$$

A is α-r.e in B if A is equal to W_e for some $e < \alpha$.

<u>2.2 DEFINITION</u> Let $B \subseteq \alpha$. The α-jump of B, denoted B', is the set

$$\{(x,e) \mid x \in W_e^B\}.$$

Clearly $\emptyset'$ is an α-r.e. set, and every α-r.e. set is α-recursive in $\emptyset'$. As usual this set is a typical example of a complete α-r.e. set. The set $\emptyset'$ is in fact of the same α-degree as the canonical $\Sigma_1(L_\alpha)$ master code studied in Chapter 1.

The following is easy to verify.

2.2 PROPOSITION

(a) If $A \leq_\alpha B$ then $A' \leq_\alpha B'$.

(b) $A <_\alpha A'$.

If A is of α-degree **a**, then the α-degree of A' is denoted **a'**. The α-degree of $\emptyset'$ is denoted $\mathbf{0}'$.

The following result, known as the Friedberg Completeness Theorem, is proved by Simpson (Simpson [1974a]).

2.3 THEOREM Let $\mathbf{b} \geq \mathbf{0}'$. Then **b** is regular if and only if there exists a regular, hyperregular α-degree **a** such that

$$\mathbf{a}' = \mathbf{a} \cup \mathbf{0}' = \mathbf{b}.$$

Before we proceed with the proof of this theorem, let us first introduce a notion which has been carefully studied in ω-recursion theory (cf. Jockusch [1981]). We will identify a set with its characteristic function. Given two α-finite functions p and q with range in 2, we say that q extends p, in symbols $p \leq q$, if $q(x) = p(x)$ for all x in the domain of p. If A is a set, and p is an α-finite function, we write $p < A$ if p agrees with the characteristic function of A on the domain of p. A string is an α-finite function with domain $\gamma < \alpha$ and range in 2. Since there is an α-recursive bijection (1.4) between α-finite sets and ordinals less than α, we call an ordinal a code for a string if they correspond to one another under this bijection. For the rest of the chapter we will reserve the symbols p, q, r etc. for strings. The length of a string p, denoted lth(p), is the least ordinal not in dom(p). Let α^* denote the $\Sigma_1(L_\alpha)$ projectum of α and let $p_1: \alpha \to \alpha^*$ be a $\Sigma_1(L_\alpha)$ projection. The α-cardinality of an α-finite set K is the least ordinal $\gamma < \alpha$ for which there is an α-finite bijection between K and γ, and this is denoted $|K|^\alpha$.

2.4 DEFINITION A set $A \subseteq \alpha$ is 1-generic if the following holds:

(a) Suppose α^* is a regular α-cardinal or $\alpha = \alpha^*$. Then for every α-finite set K, there is a $p < A$ such that either for all $e \in K$, $\text{code}(p|\gamma) \in W_e$ for some $\gamma < \text{lth}(p)$, or there is an e in K for which no extension p' of p satisfies $\text{code}(p'|\gamma) \in W_e$ for some $\gamma < \text{lth}(p')$.

(b) Suppose α^* is a singular α-cardinal. Then for every α-finite set K such that $|K|^\alpha$ is less than α^*, the conclusion of (a) holds.

It can be verified (Jockusch [1981]) that for $\alpha = \omega$, a set is 1-generic if and only if it is Cohen generic for Σ_1 sentences in the language of first order arithmetic augmented with a set variable A (for the generic set). The relevance of 1-generic sets for the problem under consideration may be seen from the following:

2.5 LEMMA Let A be 1-generic. Then

(a) A is regular and hyperregular;

(b) If the α-degree of A is **a**, then $\mathbf{a}' = \mathbf{a} \cup \mathbf{0}'$.

Proof. Let $\gamma_0 < \alpha$ be fixed. To show that $A|\gamma_0$ is α-finite, consider the α-recursive set $W = \{x \mid x \text{ is the code of a string r such that } \text{lth}(r) > \gamma_0\}$. By the 1-genericity of A, there is a $p < A$ such that either $\text{code}(p|\gamma) \in W_e$ for some $\gamma < \text{lth}(p)$, or no extension p' of p satisfies $\text{code}(p'|\gamma) \in W_e$ for some $\gamma < \text{lth}(p')$. The latter is certainly absurd, and so some $p < A$ has its code in W, i.e.

$lth(p) > \gamma_0$. Since p is α-finite, its restriction to γ_0 is still α-finite, so that $A|\gamma$ is α-finite. Hence A is regular.

Now let f be a function weakly α-recursive in A with domain $\gamma < \alpha$. We will show that the range of f is bounded below α. This will imply that A is hyperregular. For the sake of contradiction suppose that Range(f) is unbounded in α. Let κ(f) be the least ordinal $\gamma < \alpha$ such that $\sup \{f(x) | x < \gamma\}$ is unbounded in α. Suppose we have chosen the function f so that κ(f) is the minimal among all functions which exhibit the nonhyperregularity of A. We first claim that κ(f) is a regular α-cardinal. Indeed assume that this is not the case. Let $h: \delta \to \kappa(f)$ be an α-finite function mapping $\delta < \kappa(f)$ cofinally into κ(f). For each $x < \delta$, let $h^*(x) = \sup \{f(z) | z < h(x)\}$. By the choice of h, f and κ(f), the function h^* is weakly α-recursive in A and $h^{*\prime\prime}\delta$ is unbounded in α, contradicting the minimality of κ(f). Thus κ(f) is a regular α-cardinal.

Now for each $x < \kappa(f)$, let

$$W_x^* = \{code(p) | (\exists \nu)(\exists \eta)(\exists y)[\nu,\eta,x,y) \in W_e \ \& \ K_\nu \subset p \ \& \ K_\eta \subset \alpha - p]\},$$

where e is an ordinal chosen such that

$$f(x) = y \leftrightarrow (\exists \nu)(\exists \eta)[(\nu,\eta,x,y) \in W_e \ \& \ K_\nu \subset A \ \& \ K_\eta \subseteq \alpha - A]\}.$$

Now $\{W_x^*\}_{x<\kappa(f)}$ is indexed by an α-finite set K. Observe that if α^* is a singular α-cardinal, then $\kappa(f) \leq \alpha^*$ and so $|K|^\alpha$ is less than α^*. Hence by the 1-genericity of A there is a $p < A$ such that either for all $x < \kappa(f)$, $code(p|\delta)$ lies in W_x^* for some $\delta \leq lth(p)$, or there is an x for which no extension p' of p satisfies $code(p'|\delta) \in W_x^*$ for some

$\delta < \mathrm{lth}(p')$. Since f is weakly α-recursive in A, and every $x < \kappa(f)$ is in the domain of f, the latter is impossible, and so for all $x < \kappa(f)$, $\mathrm{code}(p|\delta) \in W_x^*$ for some $\delta < \mathrm{lth}(p)$. By the definition of W_x^*, it follows that Range(f) is bounded below α. This proves (a).

To prove (b), observe first of all that $\mathbf{a} \cup \mathbf{0}' \leq \mathbf{a}'$. We show that the converse also holds. Let K be an α-finite set of pairs of ordinals and suppose we wish to decide if K is contained in $\{(x,e) | x \in W_e^A\}$. Unraveling the definition of A', we wish to decide if K is contained in

$$\{(x,e) | (\exists \nu)(\exists \eta)[(\nu,\eta,x,1) \in W_e \,\&\, K_\nu \subseteq A \\ \&\, K_\eta \subseteq \alpha - A\}.$$

Now consider the α-r.e. set

$$W = \{\mathrm{code}(p) | (\forall (x,e) \in K)(\exists \nu)(\exists \eta)[(\nu,\eta,x,1) \in W_e \\ \&\, K_\nu \subseteq p \,\&\, K_\eta \subseteq \alpha - p\}.$$

As A is regular, it is possible to use $A|\gamma$, an α-finite set for each $\gamma < \alpha$, to ask if there is an extension of $A|\gamma$ whose code lies in W. This is a procedure which is α-recursive in $A \oplus \emptyset'$. As A is 1-generic, there is a $p < A$ such that either code(p) lies in W or no extension of p has its code lying in W. When such a p is found, one then knows whether $K \subseteq A'$. In a similar way one can decide if a given α-finite set is in $\alpha - A'$. Thus $\mathbf{a}' \leq \mathbf{a} \cup \mathbf{0}'$.

It follows from Lemma 2.5 that to prove Theorem 2.3 it is sufficient to construct a 1-generic set A such that $A \cup \emptyset' = B$, the given regular set lying above $\emptyset'$. This can be achieved by coding B (in the usual sense) into the set A during the construction. This is in fact not too difficult to do. The difficulty lies in the attempt to make A a 1-generic set, and as will become apparent later, this stems from the (possible) non-Σ_2-admissibility of the ordinal α. The following lemma is needed for the construction. We leave its proof to the reader.

2.6 LEMMA A function is $\Sigma_2(L_\alpha)$ if and only if it is weakly α-recursive in $\emptyset'$.

2.7 LEMMA $\sigma 2cf(\alpha) = \sigma 2cf_\alpha(\alpha^*)$.

Proof. Let $p_1: \alpha \to \alpha^*$ be a $\Sigma_1(L_\alpha)$ projection. Let $F: \sigma 2cf(\alpha) \to \alpha$ be a $\Sigma_2(L_\alpha)$ cofinality map. Then $p_1 \circ F: \sigma 2cf(\alpha) \to \alpha^*$ is $\Sigma_2(L_\alpha)$. For each $\zeta < \alpha^*$, we have by 1.20 $\text{Range}(p_1)|\zeta = K$ to be α-r.e. hence α-finite. It follows by admissibility that $(p_1^{-1})"K$ is α-finite, hence bounded below α. As a consequence $p_1 \circ F$ is unbounded in α^*. It follows that $\sigma 2cf(\alpha) \geq \sigma 2cf_\alpha(\alpha^*)$.

Conversely let $G: \sigma 2cf_\alpha(\alpha^*) \to \alpha^*$ be $\Sigma_2(L_\alpha)$ and cofinal. For each $\zeta < \alpha^*$, let $h(\zeta) = \sup\{\gamma | p_1(\gamma) < \zeta\}$. Then h is total on α^* and is $\Sigma_2(L_\alpha)$. Furthermore $h"\alpha^*$ is unbounded in α. The composition function $h \circ G$ gives a $\Sigma_2(L_\alpha)$ cofinal function from $\sigma 2cf\ (\alpha^*)$ into α. Hence $\sigma 2cf(\alpha) \leq \sigma 2cf_\alpha(\alpha^*)$.

2.8 LEMMA If $\alpha^* < \alpha$, then it is the greatest cardinal of α.

Proof. Suppose $\alpha^* < \alpha$. Let γ be such that $\alpha^* < \gamma < \alpha$. Let p_1 be as above. Then $p_1"\gamma \subseteq \alpha^*$ is α-finite, i.e. there is an α-finite injection of γ into α^*, so that γ is not an α-cardinal.

2.9 LEMMA If $\{I_\rho | \rho < \zeta\}$, $\zeta < \sigma 2cf(\alpha)$, is a simultaneous α-r.e. sequence of α-finite sets, then $\bigcup\{I_\rho | \rho < \zeta\}$ is again α-finite.

Proof. This function f which takes each $\rho < \zeta$ to $\sup I_\rho$ is $\Sigma_2(L_\alpha)$. Since $\zeta < \sigma 2cf(\alpha)$, $\sup\{f(\rho) | \rho < \zeta\}$ is bounded below α. This means that there is a stage at which every element in

$\cup \{I_\rho \mid \rho < \zeta\}$ is enumerated, so that this set must be α-finite.

<u>2.10 COROLLARY</u> If $f\colon \sigma 2cf(\alpha) \to \alpha$ is $\Sigma_2(L_\alpha)$, then

(a) $\{f(\rho) \mid \rho < \zeta\}$, $\zeta < \sigma 2cf(\alpha)$, is α-finite.

(b) The sequence $\{f(\rho) \mid \rho < \sigma 2cf(\alpha)\}$ is $\Delta_2(L_\alpha)$.

Proof. Let $f(\rho) = \lim_\sigma f'(\sigma,\rho)$, where f' is α-recursive (Theorem 1.50). Then

$$I_\rho = \{\sigma \mid (\forall \nu<\sigma)(\exists \tau)[\nu < \tau < \sigma \;\&\; f'(\nu,\rho) \neq f'(\tau,\rho)\}$$

is α-finite for each ρ and the hypothesis of Lemma 2.8 is satisfied. It follows that $\cup \{I_\rho \mid \rho < \zeta\}$ is α-finite. Then the set $\{i_\rho \mid i_\rho$ = the least ordinal not in $I_\rho\}$ is α-finite. But $f(\rho) = f'(i_\rho,\rho)$. This proves (a). As the sequence $\{f'(i_\rho,\rho) \mid \rho < \sigma 2cf(\alpha)\}$ is $\Delta_2(L_\alpha)$, we have (b).

We are now ready to prove Theorem 2.3. By Theorem 1.22 we know that $\emptyset'$ is regular. One may show in a similar way that if A is regular and hyperregular, then so is A'. We focus our attention then on the other direction of the theorem. Let F be a $\Sigma_2(L_\alpha)$ cofinal function from $\sigma 2cf(\alpha)$ into α. Also let G be a $\Sigma_2(L_\alpha)$ cofinal function from $\sigma 2cf_\alpha(\alpha^*) = \sigma 2cf(\alpha)$ into α^*. Let B be a regular set in which $\emptyset'$ is α-recursive. We will produce a 1-generic set A by stages such that $A = \{q_\zeta \mid \zeta < \sigma 2cf(\alpha)\}$, where $q_\zeta > q_\mu$ if $\zeta > \mu$, and $q_{2\zeta+1} = q_{2\zeta} * B|F(\zeta)$ (as in classical recursion theory, we may define the concatenation $p*q$ of two strings p and q to be the string

$$p \cup \{(\mathrm{lth}(p)+\xi, q(\xi)) \mid \xi < \mathrm{lth}(q)\}).$$

We further require that the sequence $\langle q_\zeta \mid \zeta < \sigma 2cf(\alpha)\rangle$ be α-recursive in B and in $A \oplus \emptyset'$. The former ensures that $A' \leq B$ while the latter $B \leq A'$ (by Lemma 2.5, $A' \equiv_\alpha A \oplus \emptyset'$).

Finally by redefining G if necessary, we may assume that $G"\sigma 2cf_\alpha(\alpha^*)$ is contained in $(p_1)"\alpha$, and as in the proof of Lemma 2.7 one may choose G so that $F = p_1^{-1} \circ G$.

Case 1. α^* is either a regular α-cardinal or equal to α.

Suppose that $q_{2\nu}$ is defined with the property that for all $\rho < G(\nu)$, either:

For all $e \in K_{p_1^{-1}(\rho)}$, $\mathrm{code}(q_{2\nu}|\gamma) \in W_e$ for some $\gamma < \ell th(q_{2\nu})$, or there is an $e \in K_{p_1^{-1}(\rho)}$ for which there is no extension p' of $q_{2\nu}$ with code $(p'|\gamma) \in W_e$ for some $\gamma < \mathrm{lth}(p')$.

Recall that $K_{p_1^{-1}(\rho)}$ is the $p_1^{-1}(\rho)$th α-finite set. We will define $q_{2\nu+2}$ so that it enjoys the same property as $q_{2\nu}$ when $G(\nu)$ above is replaced by $G(\nu+1)$. Notice first of all that the set of all ordinals $\rho < G(\nu+1)$ in the image of p_1 is α-finite. We denote this set by K. If $p_1(e) = \rho \in K$, let K_e be denoted K_ρ^*. Thus $\{K_\rho^*\}$, $\rho \in K$, is an α-finite collection of α-finite sets.

Now by Lemma 2.8, every K_ρ^* may be identified (α-recursively) with a subset of α^*. We will do this and henceforth assume that $K_\rho^* \subseteq \alpha^*$ for all $\rho \in K \subseteq G(\nu+1) < \alpha^*$. Given a string p and $\rho \in K$, let $e(p,\rho)$ be the least $e \in K_\rho^*$, if exists, for which $\mathrm{code}(p|\gamma) \notin W_e^{\mathrm{lth}(\rho)}$ for any $\gamma < \mathrm{lth}(p)$. Otherwise let it be 0. Then for each p, $e(p,\rho)$ may be computed α-recursively. We now define a sequence of strings $\{p(t)\}$:

Let $p(0) = q_{2\nu+1}$. If $p(t)$ is defined, let $p(t+1)$ be the least string p extending $p(t)$ such that for some ρ in K, $e(p,\rho) > e(p(t), \rho)$. If t is a limit ordinal and $p(u)$ is defined for all $u < t$, let $p(t) = \bigcup p(u)$, $u < t$. Let t_0 be the least stage where $p(t)$ is not defined. We claim that $t_0 < \alpha$. By way of contradiction, assume that $t_0 = \alpha$.

First suppose that α^* is a regular α-cardinal. Let $\rho(t+1)$ be the least ordinal ρ in K such that at stage t+1, $e(p(t+1),\rho) > e(p(t),\rho)$. Let $I_\rho = \{t \mid \rho(t+1) = \rho\}$. Then $\{I_\rho \mid \rho \in K\}$ is a simultaneous α-r.e. sequence of α-finite sets. Furthermore by our assumption that $K^*_\rho \subseteq \alpha^*$ for each ρ in K, we see that I_ρ has order type at most α^*. The set H of ρ's in K for which order type of I_ρ is α^* is $\Sigma_1(L_\alpha)$, hence α-finite by the definition of α^*. Let γ_0 be the least ordinal greater than each element of I_ρ, for ρ in H. Clearly $\gamma_0 < \alpha$. Now those ordinals ρ in K for which order type of I_ρ is less than α^* is also α-finite. Call this set H*. For each ρ in H*, let $J_\rho = I_\rho \cap \{\delta \mid \gamma_0 \leq \delta \leq \gamma+\alpha^*\}$. We have a simultaneous α-r.e. sequence $\{J_\rho \mid \rho \in H^*\}$ of α-finite sets each having order type less than α^* Furthermore, $\cup \{J_p \mid \rho \in H^*\} = \{\delta \mid \gamma \leq \delta \leq \gamma_0+\alpha^*\}$. Since $|H^*|^\alpha < \alpha^*$, this contradicts the assumption that α^* is a regular α-cardinal.

It follows that the stage t_0 is less than α. It is clear that this is a limit ordinal. Now let $q_{2\nu+2} = \cup\, p(t)$, $t < t_0$. It can be verified that this string satisfies our requirement.

Now suppose that $\alpha^* = \alpha$. The set $H = \{(e,\rho) \mid e \in K^*_\rho,\ \rho \in K\}$ is α-finite. Define $\rho(t+1)$ as in the previous case. Then

$$M = \{(\varepsilon,\rho) \mid (\exists t)[\rho = \rho(t+1)\ \&\ e = e(\rho(t+1),\ \rho(t+1))]\}$$

is a $\Sigma_1(L_\alpha)$ set bounded below α, hence α-finite. Thus t_0 is the least stage u where for all (e,ρ) in M, there is a $t < u$ such that $\rho = \rho(t+1)$ and $e = e(p(t+1),\ \rho(t+1))$. Since this stage can be located effectively, we have $t_0 < \alpha$. We now define $q_{2\nu+2}$ as in Case 1.

Case 2. α^* is a singular α-cardinal.

Define K as in Case 1. For each $\rho < G(\nu+1)$, define (where possible) K^*_ρ as before. Again we may regard K^*_ρ to be a subset of α^*.

We will now consider only those ρ's in K where $|K_\rho^*|^\alpha < |G(\nu+1)|^\alpha$. For convenience let us again denote this set by K. It is then possible to assume that each K_ρ^* is a subset of $G(\nu+1)$. We define our sequence $\langle q_\nu \rangle$ in the following way.

Suppose that $q_{2\nu}$ is defined (hence so is $q_{2\nu+1}$), where $\nu < \sigma 2cf(\alpha^*)$. Set $p(0) = q_{2\nu+1}$. Assume that $p(t)$ is defined. For $\rho < G(\nu+1)$, let $e(p(t),\rho)$ be the least $e \in K_\rho^* \subset G(\nu+1)$ such that $code(p|\gamma) \notin W_e^{lth(p)}$ for all $\gamma < lth(p)$. Set $p(t+1)$ to be the least string p extending $p(t)$ such that $e(p,\rho) > e(p(t),\rho)$ for some ρ in K. If t is a limit ordinal and $p(u)$ is defined for all $u < t$, let $p(t) = \bigcup p(u)$, $u < t$. Let t_0 be the least ordinal t for which $p(t)$ is not defined. Once again we claim that $t_0 < \alpha$.

Define I_ρ as in Case 1. The sequence $\langle I_\rho \rangle$ is a simultaneous α-r.e. collection of α-finite sets, each having order type at most $G(\nu+1)$. If $t_0 = \alpha$, then clearly $\bigcup \{I_\rho | \rho \in K\} = \alpha$. The function $i(e,\rho)$ = the eth element of I_ρ is then an α-recursive bijection between a subset of $G(\nu+1) \times G(\nu+1)$ and α. This is not possible by 1.20. Hence $t_0 < \alpha$.

Now let $q_{2\nu+2} = \bigcup p(t)$, $t < t_0$.

This completes the construction of the sequence $\langle q_\nu)$ at successor stages.

Now suppose that q_ν, $\nu < \lambda$, is defined and λ is a limit ordinal. The sequence $\langle q_\nu | \nu < \lambda \rangle$ is $\Sigma_2(L_\alpha)$ (since only α-finitely much information about B, namely $B|F(\lambda)$, is used), and hence α-finite by 2.10 (a). One may then set $q_\lambda = \bigcup q(\nu)$, $\nu < \lambda$.

We let $A = \bigcup q_\nu$, $\nu < \sigma 2cf(\alpha)$. We claim that A is 1-generic. The claim is clear if either α^* is a regular α-cardinal or if $\alpha = \alpha^*$. Suppose that α^* is a singular α-cardinal. Let N be an α-finite set such that $|N|^\alpha < \alpha^*$. As in the construction we may regard $N \subseteq \alpha^*$.

Then there is $\nu < \sigma 2cf_\alpha(\alpha^*)$ such that N has α-cardinality less than that of $G(\nu)$. Suppose we have chosen the least such ν. The construction then ensures that either for each $e \in N$ there is a $\gamma < lth(q_{2\nu})$ such that code $(q_{2\nu}|\gamma) \in W_e$, or there is an $e \in N$ for which no extension q of $q_{2\nu}$ satisfies code$(q|\gamma) \in W_e$, for all $\gamma < lth(q)$. Thus A is 1-generic.

By Lemma 2.5(b) $A' \equiv_\alpha B \oplus \emptyset'$. We now show that $A \oplus \emptyset'$ is of the same α-degree as that of $B \equiv_\alpha B \oplus \emptyset'$.

Claim For each $\nu < \sigma 2cf(\alpha)$, let $Q(\nu)$ be the sequence $\langle q_\zeta | \zeta < 2\nu \rangle$. Then the function Q is weakly α-recursive in B and in $A \oplus \emptyset'$.

Fix ν. To compute $Q(\nu)$ use $\emptyset'$ to locate $f(\nu)$ (by Lemma 2.6). By Corollary 2.10 the set $\{F(\zeta)|\zeta < \nu\}$ is α-finite. Further more the regularity of B implies that $B|\gamma$, where $\gamma = \sup \{F(\zeta)|\zeta < \nu\}$ is α-finite. One can then use the α-finite information and $\emptyset'$ to obtain $Q(\nu)$. Observe now to get $q_{2\zeta+1}$ from $q_{2\zeta}$ one simply appeals to $B|\gamma$. To obtain $q_{2\zeta+2}$ from $q_{2\zeta+1}$ one uses $\emptyset' \leq_\alpha B$ as discussed below.

Consider all α-finite sequences T of strings $\langle p_\zeta | \zeta < 2\nu \rangle$ such that $\cup \langle p_\zeta | \zeta < 2\nu \rangle < A$, with the extra condition that $lth(p_{2\zeta+1}) = lth(p_{2\zeta}) + F(\zeta)$. These are the candidates for strings in $Q(\nu)$. One examines all sequences $\langle p_\zeta | \zeta < 2\nu \rangle$ looking for the least one (in a sense to be given presently) having the property that $p_{2\zeta+1} = q_{2\zeta+1}$ and $p_{2\zeta+2}$ is obtained from $p_{2\zeta+1}$ by the procedure detailed in the construction of $\langle p(t)|t < t_o \rangle$ given earlier. To do this one needs the help of $\emptyset'$. Specifically using $\emptyset'$ one goes through sequences in T in stages. At stage s, say, one will take s steps to examine the first s sequences in T, using $\emptyset'$. In this way one gets α-finite information about sequences in T at the end of stage s. By the construction, however, there is a sequence $\langle p_\zeta | \zeta < 2\nu \rangle$ which agrees with $q_{2\zeta+2}$, and the least one obtained under our process of examination takes place before one reaches stage α.

Thus the computation of $q_{2\zeta+2}$ from $q_{2\zeta+1}$ may be obtained from $\emptyset'$. It follows that $Q(\nu)$ may be computed from B since $\emptyset \leq_\alpha B$. Hence $Q \leq_{\omega\alpha} B$.

To compute $Q(\nu)$ from $A \oplus \emptyset'$, note again by Lemma 2.6 that $\langle F(\zeta)|\zeta < \nu\rangle$ and $\langle G(\zeta)|\zeta < \nu\rangle$ may be obtained from $\emptyset'$. Using this one may obtained $q_{2\zeta+1}$ from $q_{2\zeta}$ using A. since $q_{2\zeta+2}$ may be obtained from $q_{2\zeta+1}$ using $\emptyset'$, we have as a consequence $Q \leq_{\omega\alpha} A \oplus \emptyset'$.

Clearly both B and $A \oplus \emptyset'$ are α-recursive in the full sequence $\langle Q(\nu)|\nu < \sigma 2cf(\alpha)\rangle$. It follows that $B \equiv_\alpha A \oplus \emptyset'$, and this completes the proof of Theorem 2.3.

Theorem 2.3 implies that the following two statements are equivalent. Let $\mathbf{b} \geq \mathbf{0}'$.

(a) Every degree greater than or equal to **b** is the jump of a regular, hyperregular degree (i.e. an α-degree which contains a regular, hyperregular set).

(b) Every degree greater than or equal to **b** is regular.

The following question is then natural to ask: For which admissible ordinals α do there exist an α-degree **b** such that every $\mathbf{c} \geq \mathbf{b}$ is regular ? Assuming the Generalized Continuum Hypothesis, Simpson [1974a] has shown that if $|\alpha|$ (the 'real cardinality' of α) is regular, then α enjoys this property if and only if every subset of cardinality less than $|\alpha|$ is α-finite. In particular, every countable admissible ordinal α contains an α-degree **b** for which every $\mathbf{c} \geq \mathbf{b}$ is regular. The situation where $|\alpha|$ is singular is not known (cf. Simpson [1974]).

CHAPTER 3

THE α-FINITE INJURY METHOD

Starting with this chapter we embark upon the study of the priority method in admissible recursion theory. Since the method is considered by many to be the hallmark of recursion theory, the emphasis given in this monograph on this aspect of recursion theory is not without justification. We will only deal with α-finite injury method in this chapter, and defer the consideration of α-infinte injury till later chapters. Even in the realm of α-finite injury, there are two broad categories into which all arguments fall: the Friedberg-Muchnik type (the bounded type) and the Sacks splitting type (the unbounded type). The prototype of each of these is best exemplified respectively in the solution of Post's problem for all admissible ordinals (Sacks and Simpson [1972]) and in the proof of the Sacks Splitting Theorem (Shore (1975a]).

Section 1: Post's Problem

Post [1944] asked if there exist two recursively enumerable sets which are Turing incomparable. The solution of Post's Problem by Friedberg [1957a] and Muchnik [1958] was historically the beginning of the method of priority. The search for the correct generalization of the Friedberg-Muchnik Theorem and the introduction of finite injury method for admissible ordinals was not a smooth-sailing one. It was of course closely linked to the search for a 'correct' generalized recursion theory. One may indeed be tempted to define a 'good' recursion theory to be one in which Post's Problem is true. This implies the necessity for the introduction of correct notions of finiteness, effectiveness, and reducibility. Thus for example for $\alpha = \omega_1^{CK}$, the suitability of adopting hyperdegrees as the generalization of Turing degrees was rejected because each Π_1^1 set of integers (considered to be a good candidate for α-r.e. sets) occupies only one of two hyperdegrees: the degree of the hyperarithmetical sets and the degree of Kleene's $\mathcal{O}$. Post's Problem therefore has a negative solution in this situation. The fact that the problem has a positive solution for all admissible ordinals in terms of α-degrees stimulated a surge of active investigations in α-recursion theory beginning in the early 1970s.

There are two approaches to the solution of Post's Problem. The first one is in the original solution of Sacks and Simpson and has a strong set-theoretic and model-theoretic flavor. The second, based on the idea of the tame $\Sigma_2(L_\alpha)$ projectum, was introduced by Lerman [1972] and is both more recursion-theoretic and intuitive. However, as was pointed out in Lerman [1972], the Sacks-Simpson approach has the advantage that one may locate in certain cases bounds on the stages where initial segments of requirements cease to be injured. This provides a good insight into how ordinals below α act to control the behavior of various requirements. We will present here the Sacks-Simpson proof of the theorem, the purpose being to bring forth the point that in higher recursion theory, model-theoretic and set-theoretic ideas are both important and useful. One should also mention, in this connection, that Lerman (unpublished) has given a solution of Post's problem uniform for all admissible ordinals.

Recall that if K is α-finite, then $|K|^\alpha$ denotes the α-cardinality of K. The next Lemma is a useful counting result for α-finite argument.

<u>3.1 LEMMA</u> Let κ be a regular α-cardinal. Suppose that

$\langle I_\varepsilon | \varepsilon < \nu \rangle$, $\nu < \kappa$, is a simultaneously α-r.e. sequence of α-finite sets, with the property that $|I_\varepsilon|^\alpha < \kappa$ for each $\varepsilon < \nu$. Then $\bigcup I_\varepsilon$, $\varepsilon < \nu$, is α-finite and has α-cardinality less than κ.

Proof. Let f be a one-one α-recursive function with domain $\beta < \alpha$ and range $\bigcup I_\varepsilon \times \{\varepsilon\}$, $\varepsilon < \nu$. We claim that $\beta < \kappa$. Suppose this is false. Then $\kappa \subseteq \mathrm{dom}(f)$ and the restriction of f to κ is an α-recursive function f* such that $(f^*)''\kappa = K = \bigcup J_\varepsilon \times \{\varepsilon\}$, $\varepsilon < \nu$, with $J_\varepsilon \subseteq I_\varepsilon$ for each ε. Thus $|J_\varepsilon|^\alpha < \kappa$ and so by the regularity of κ the α-cardinality of K is less than κ. But f* is one-one, and so $|K|^\alpha = \kappa$. This is a contradiction and so $\mathrm{dom}(f) < \kappa$. The conclusion of the lemma now follows.

<u>3.2 DEFINITION</u> The ordinal $\beta < \alpha$ is α-stable if L_β is a Σ_1 elementary substructure of L_α.

As in Proposition 1.9, there is an effective enumeration of all partial α-recursive functions. We denote these by $\{f_\gamma | \gamma < \alpha\}$.

<u>3.3 LEMMA</u> Suppose that $\alpha = \alpha^*$ and κ_0 is the greatest cardinal of α. Let δ be such that $|\delta|^\alpha = \kappa_0$. Let

$$H_\delta = \{f_\gamma(x_1,\ldots,x_n) | n < \omega \ \&\ \gamma,\ x_1,\ldots,x_n < \delta\}.$$

Then H_δ is both an ordinal and α-finite.

Proof. We first show that H_δ is an ordinal. Let σ be in H_δ. Then there is an α-finite surjection of κ_0 onto σ. Choose the least one in the α-recursive enumeration of α-finite sets. Let this be K_γ. Then γ is in H_δ and so $K_\gamma''\kappa_0 = \sigma \subseteq H_\delta$. Hence H is an ordinal.

If $H_\delta = \alpha$, then since H_δ is the "Σ_1-Skolem Hull" of γ in α, one obtains from the definition of H_δ a partial α-recursive function with domain contained in γ^ω (ordinal exponentiation) and range all of α. But clearly $\mathrm{dom}(H_\delta)$ is α-r.e. and bounded below α. Since α

$= \alpha^*$, it is α-finite and so by admissibility $H_\delta < \alpha$, hence α-finite.

The following corollary is now immediate.

3.4 COROLLARY If $\alpha = \alpha^*$ and there is a largest α-cardinal, then the set of α-stable ordinals is unbounded in α.

The proof of Lemma 3.3 also shows that every α-cardinal κ is α-stable, for if $\delta < \kappa$, then $|H_\delta|^\alpha < \kappa$ so that $H_\delta < \kappa$. The solution of Post's problem for admissible ordinals is divided into cases, depending on the position of α^* in relation to α. Let us first state the theorem to be proved (Sacks and Simpson [1972]).

3.5 THEOREM There exists two α-r.e. sets A and B such that $A \not\leq_{w\alpha} B$ and $B \not\leq_{w\alpha} A$.

Observe firstly that since we have an α-recursive projection $p_1: \alpha \to \alpha^*$, it is in fact possible to index all α-r.e. sets and α-finite sets using α^* (instead of α). Thus we may for example let the eth α-r.e. set W_e be indexed by $p_1(e)$ instead of by e. This can similarly be done for α-finite sets. The advantage of doing this is that in the case when $\alpha^* < \alpha$, it provides one with a shorter indexing list of the requirements needed to carry out α-finite priority arguments over arbitrary admissible ordinals. Indeed as the reader will soon observe, a short indexing list suitable for a particular priority argument is quite often the key to the successful solution of a problem in α-recursion theory.

Two α-r.e. sets A and B will be constructed in stages. The objective in the construction is to achieve the following: For all e $< \alpha^*$,

R_e: $\{e\}^B \neq A$ and

S_e: $\{e\}^A \neq B$.

R_e and S_e will be called requirements. As in the case for ω, different requirements tend to conflict with each other. A system of priorities is introduced to resolve this. At any stage of the construction, an attempt will be made to satisfy the requirement T of highest priority which is not satisfied. Certain ordinals will be enumerated into either the set A or B, with the precaution that ordinals which are specifically excluded by requirements of higher priority than that of T should remain excluded. Now in the process of satisfying requirement T, certain requirements of lower priority which were previously satisfied may be 'injured', in the sense that the ordinals specifically excluded by them may enter either A or B, as the case may be, at this stage. In classical recursion theory, this situation is taken care of by the observation that each requirement gets injured only finitely many times, so that the first n requirements will stop injuring requirements of lower priority after a finite stage, allowing the (n+1)th requirement to have a completely open field in which to operate, thus providing plenty of opportunities for it to be satisfied. Two important features are used to achieve this: the regularity of the ordinal ω and the recursiveness of the priority system. These two features may still be put to good use whenever α is an infinite regular cardinal. But in general one has to resort to a more sophisticated analysis of the admissible ordinal in question and introduce a priority system that is quite possibly non-α-recursive. In addition to these, the construction of the α-r.e. set may require a new twist. The solution of Post's problem provides perhaps the simplest illustration of how this is done.

We split the proof of Theorem 3.5 into two cases:

<u>Case 1:</u> $\alpha^* < \alpha$. Given the $\Sigma_1(L_\alpha)$ projection $p_1\colon \alpha \to \alpha^*$, we

say that requirement R_d has higher priority than requirement R_e if $p_1(d) < p_1(e)$. By 1.20 the set of ordinals less than $p_1(e)$ in the image of p_1 is α-finite. It follows that p_1^{-1} on the initial segment $p_1(e)$ is α-finite and so there are only α-fintely many requirements of the type R having higher priority than that of R_e. If $p_1(d) < p_1(e)$, we also say that S_d has higher priority than R_e. On the other hand we agree that R_e has higher priority than S_e (the same e). We then obtain an α-recursive priority system, in the sense that one may effectively decide, given two arbitrary requirements, which one has the higher priority. Note that given e, it is not possible to determine in an a priori way the set of requirements occupying a higher priority than that of R_e or S_e. At stage σ, however, it is possible to effectively determine the set of requirements R_d and S_d, $d < \sigma$, of higher priority than that of R_e and S_e. Observe that for any two given requirements, if at stage σ it is known that one has higher priority than the other, then this fact remains true at all subsequent stages.

The sets A and B will be constructed in stages. At the end of stage σ one obtains α-finite sets A^σ and B^σ. At the beginning of stage σ, we set $A^{<\sigma} = \bigcup A^\tau$, $\tau < \sigma$ and $B^{<\sigma} = \bigcup B^\tau$, $\tau < \sigma$. We set $A = \bigcup A^\sigma$ and $B = \bigcup B^\sigma$, $\sigma < \alpha$. We will take care of requirements of type R_e during even stages, and those of type S_e during odd stages. Since these two situations are symmetric, we will elaborate only on the construction at even stages. We say that

$$(3.6)\quad \{e\}_\sigma^{A^{<\sigma}}(x) \text{ is defined}$$

If there exist $K_\nu \subseteq A^{<\sigma}$ and $K_\eta \subseteq \alpha - A^{<\sigma}$ such that $(\nu,\eta,x,y) \in W_e^\rho$, where ρ is $(\sup A^{<\sigma})+1$ and $y = 0$ or 1.

The requirement R_e is said to be satisfied at the beginning of stage σ if there is an x such that

(3.7) $\{e\}_{\sigma}^{B<\sigma}(x) \neq A^{<\sigma}(x)$,

where the left hand side of 3.7 is defined and where $A^{<\sigma}(x) = 1$ if x is in $A^{<\sigma}$, and equals 0 otherwise. If (3.7) holds for x then we say that x is a witness to the satisfaction of requirement R_e. The least pair (K_ν, K_η) which appears in the computation of the left hand side of 3.7 (where $K_\nu \subseteq A^{<\gamma}$ and $K_\eta \subseteq (\text{Sup } A^{<\alpha})+1 - A^{<\sigma}$) will be called the pair of α-finite sets used in the computation of the witness to the satisfaction of requirement R_e). We define these notions similarly for requirements S_e. From now on we let T be either R or S.

The construction applies the technique of movable markers. Each requirement is given a marker which at any stage of the construction occupies a certain position in α. We denote this position by $m(e,R,\sigma)$ for requirement R_e at stage σ. (Also $m(e,S,\sigma)$ refers to the position of the marker for S_e at stage σ). The marker position of each requirement may not go down (imagine the ordinal α to be a column erected vertically so that 0 is at the base of the column). It may however go up in the course of the construction. We ensure, by induction, that at the beginning of stage σ, if R_e (or S_e) is not satisfied, then $\lim_{\tau<\sigma} m(e,R,\tau)$ (if exists) is not in $A^{<\sigma}$ (similarly for $\lim_{\tau<\sigma} m(e,S,\tau)$). If the limit exists, set $m(e,R,\gamma)$ to be the limit; otherwise let $m(e,R,\sigma)$ to be the least ordinal ζ with the following properties:

(3.9) ζ is not in $A^{<\sigma}$

(3.10) $\zeta > m(d,T,\sigma)$ for every T_d of higher priority at stage σ

(3.11) $\zeta >$ the supremum of ordinals occuring in an α-finite pair that was used earlier in a computation to satisfy some requirement of higher priority which remains satisfied at the beginning of stage σ.

If R_e is satisfied at the beginning of stage σ, set $m(e,R,\sigma)$ to be the limit of $m(e,R,\tau)$, $\tau < \sigma$. It will be seen from the construction that the limit exists in this case. Again do this similarly for requirement S_e. The point is that $m(e,R,\sigma)$ is a potential candidate for a witness to the satisfaction of R_e. Hence it is most convenient to let it be an ordinal which does not interfere with the computations used to satisfy requirements of higher priority. In this way the 'good' computations obtained earlier will be preserved, while there is a possibility that if $m(e,R,\sigma)$ turns out to be a witness, the computation used to obtain this fact will not be destroyed.

Suppose now σ is even. Find the requirement R_e, $e < \sigma$ of highest priority at stage σ, such that R_e is not satisfied at the beginning of stage σ and

(3.12) $\{e\}_\sigma^{B^{<\sigma}}(m(e,R,\sigma))$ is defined and not equal to $A^{<\sigma}(m(e,R,\sigma))$ (recall that by induction $m(e,R,\sigma) \notin A^{<\sigma}$);

We set $A^\sigma = A^{<\sigma} \cup \{m(e,R,\sigma)\}$ if the output of the computation $\{e\}_\sigma^{B^{<\sigma}}(m(e,R,\sigma))$ is equal to 0, otherwise set $A^{<\sigma} = A^\sigma$. Set $B^{<\sigma} = B^\sigma$. This completes the construction at even stage σ. We reverse the roles of A and B at odd stages.

The requirement R_e is said to be 'injured' at stage σ if there is a $\rho < \sigma$ such that R_e is satisfied at the beginning of each stage $\rho \leq \tau < \sigma$, and R_e is not satisfied at the beginning of stage $\sigma+1$ (or equivalently at the end of stage σ when A^σ is specified). Define this similarly for requirement S_e. Suppose $p_1(e) = \varepsilon$. Let $I_\varepsilon(R)$ (and respectively $I_\varepsilon(S)$) be the set of stages γ where $R_e(S_e)$ gets injured.

Clearly $I_\varepsilon(R)$ and $I_\varepsilon(S)$ are α-recursive for each e. We arrange $I_\varepsilon(T)$ according to the same priority as we gave to the requirements R_e and S_e.

Let ot(e) be the order type of

$$\{\delta \mid \delta < p_1(e) \ \& \ \delta \text{ in Range}(p_1)\}.$$

<u>3.13 LEMMA</u> Let $p_1(e) = \varepsilon$. If σ is a stage where $I_\delta(T) \subseteq \sigma$ for every $I_\delta(T)$ of higher priority than that of $I_\varepsilon(R)$, then there are at most ot(e) many ordinals in $I_\varepsilon(R)$ greater than σ.

Proof. This is a consequence of the following observation: the requirement R_e gets injured at stage $\rho > \sigma$ only if an attempt is made to satisfy a requirement of higher priority. By assumption, after stage σ no such attempt will ever injure any requirement of higher priority than that of R_e. This means that those of higher priority than that of R_e, if they are satisfied at stage σ, will remain satisfied thereafter. It follows that the stages greater than σ where a requirement of higher priority is satisfied include the stages after σ where R_e is injured. But there are at most order type ot(e) many stages of the former type, perhaps one for each requirement of higher priority. Hence R_e may be injured at most ot(e) many times after σ.

Observe that for each e,

$$\langle I_\zeta(T) \mid p_1(d) = \zeta \ \& \ T_d \text{ of higher priority than } R_e \rangle$$

is a simultaneous α-r.e. sequence of α-recursive sets. Indeed one can prove much more:

<u>3.14 LEMMA</u> let κ be an α-cardinal. If $\bigcup I_\zeta(T)$, where $p_1(d) = \zeta$ and T_d of higher priority than R_e, is α-finite and has α-cardinality less than κ, then the same conclusion holds for $I_\varepsilon(R)$ as

well.

Proof. This follows from 3.13.

3.15 LEMMA Let κ be a regular α-cardinal. Suppose that each $I_\zeta(T)$, where T_ζ is of higher priority than R_e, is α-finite of α-cardinality less than κ. Then the union of these $I_\zeta(T)$'s is α-finite and of α-cardinality less than α.

Proof. This follows from Lemma 3.1.

3.16 LEMMA For all T and d, there is a stage σ_o where $m(d,T,\sigma_o) = m(d,T,\sigma)$ for all $\sigma > \sigma_o$.

Proof. From the construction and the definition of the function m it follows that the marker for T moves its position only when requirement T_d gets injured by a requirement of higher priority. Lemmas 3.14 and 3.15 then ensure that every marker stabilizies after some stage.

3.17 LEMMA For all e, if $\{e\}^B$ is a total function, then there is a stage σ where R_e is satisfied thereafter using $m(e,R,\sigma)$ as the witness.

Proof. Since $\alpha^* < \alpha$, by Lemma 2.8 it is the greatest α-cardinal. It is then either a regular α-cardinal or a limit of regular α-cardinals. In any case there is a regular α-cardinal $\kappa > e$. Combining 3.14 and 3.15 we know that there is a stage σ where $\cup I_\delta(T)$, $P_1(d) = \delta$ and of higher priority than R_e, is a subset of σ. Suppose that τ is chosen to be the least such ordinal. Then at the beginning of stage τ, if R_e is satisfied it will be satisfied at all subsequent stages, since no requirement of higher priority may injured this computation. On the other hand, if R_e is not satisfied at the beginning of stage τ, consider the ordinal $m(e,R,\tau)$. By 3.17 and its proof, since R_e does not get injured by a requirement of higher priority anymore, the marker assigned to R_e has reached its final position, namely $m(e,R,\tau)$. By our choice this ordinal is not in $A^{<\gamma}$. Since $\{e\}^B(m(e,R,\tau))$ is defined, there is a stage σ where $\{e\}_\sigma^{B^{<\sigma}}(m(e,R,\tau))$ is defined. Now the construction keeps the ordinal $m(e,R,\tau)$ out until such time as $\{e\}_\sigma^{B^{<\sigma}}(m(e,R,\tau))$ is defined, at which stage steps are taken to ensure that the requirement R_e is satisfied. This proves the lemma.

The proof of Theorem 3.5 for the case $\alpha^* < \alpha$ is complete.

Case 2: $\alpha^* = \alpha$. Two situations may arise in this case. The first one is when α is the limit of α-cardinals. Then α is in fact the limit of regular α-cardinals, and so one may construct the sets A, B in the same manner as in Case 1, with practically the same proof (it is helpful to observe in this case that the function p_1 is just the identity function). We will therefore only consider the situation when there is a greatest α-cardinal $\kappa_0 < \alpha$.

By 3.4, the set St of α-stable ordinals is unbounded in α. There are two possibilities: either St has order type less than α or equal to α. We start with the first possibility. Observe that St is $\Pi_1(L_\alpha)$.

Let $\lambda < \alpha$ be the order type of St. We define a $\Sigma_2(L_\alpha)$ map from $\kappa_0 . \lambda$ onto α. For each $z < \alpha$, let $\rho(z)$ be the least ordinal ρ such that K_ρ is an α-finite function from κ_0 onto z. For each $\sigma < \alpha$, let $St(\sigma)$ be the set of σ-stable ordinals and let $\lambda(\sigma)$ be its order type. Clearly $St(\sigma)$ is α-finite for each σ. Let $b(\sigma,\xi)$ be the ξ member of $St(\sigma)$ (hence $\xi < \lambda(\sigma)$). For $\varepsilon = \kappa_0 . \xi + \gamma$, where $\xi < \lambda$ and $\gamma < \kappa_0$, let

$$k'(\sigma,\varepsilon) = K_{\rho(b(\sigma,\xi))}(\gamma).$$

Then k' is α-recursive and $\lim_\sigma k'(\sigma,\varepsilon) = k(\varepsilon)$ exists for each $\varepsilon < \kappa_0.\lambda$. Indeed the reader may verify that this follows from

(3.18) $b(\sigma,\xi) = \beta_\xi$, the ξth α-stable ordinal, for all $\xi < \lambda$. Furthermore for each $\varepsilon < \kappa_0.\xi$, there is a $\sigma_\varepsilon < \beta_\xi$ such that $k'(\sigma,\delta) = k(\delta)$ for all $\sigma > \sigma_\varepsilon$ and $\delta < \varepsilon$. In other words, proper initial segment of β_ξ attains its final value before stage β_ξ.

The function k is a surjection from $\kappa_0.\lambda$ onto α. We define requirements R_e and S_e as before. The function k provides us with a system of priority which is Σ_2 over L_α. From now on the ordianl ε will range over $\kappa_0.\lambda$. If ε is the least ordinal such that $k(\varepsilon) = e$, then the priority of requirement T_e is ε (where T is again either R or S). For reason of convenience, we stipulate that R_e has higher priority than S_e (for the same e). Thus if $k(\delta) = d < k(\varepsilon) = e$, and δ, ε are the respectively the least ordinals mapped onto d and e, then T_d has higher priority than T_e (T_d and T_e may be of different types, i.e they do not have to be both R or both S). The function k' provides an approximation to the 'real' priority of T. Thus if ε is the least ordinal such that $k'(\sigma,\varepsilon) = e$, then we say that the priority of T_e at stage σ is ε.

The notions of satisfaction, use and injury are as in Casel. Again we concentrate only on the even stage of the construction. Define the marker position of requirement T_e at stage σ as follows: Suppose ε is the least ordinal such that $k'(\sigma,\varepsilon) = e$. If R_e is satisfied at the beginning of stage σ, let $m(e,R,\sigma) = \lim_\tau m(e,R,\tau)$, $\tau < \sigma$. Otherwise let it be the least ordinal meeting conditions (3.9), (3.10), and (3.11). Define $A^{<\sigma}$ in the same way as in Case 1, keeping in mind that the priority system is now given by the function k and its approximation k' at stage σ. We now verify that the sets A and B have incomparable α-degrees. Let $I_\varepsilon(T)$ be the set of stages where T_e gets injured.

<u>3.19 LEMMA</u> Fix $\beta_\xi < \alpha$. Suppose that $\bigcup I_\varepsilon(T)$, $\varepsilon < \kappa_0.\lambda$, is a subset of β_ξ. Then the same conclusion holds for $\kappa_0.\xi$ replaced by $\kappa_0.(\xi+1)$.

Proof. We prove the lemma by induction. If $\bigcup I_\varepsilon(T)$, $\varepsilon < \varepsilon'$, is a subset of $\beta_{\xi+1}$, where $\varepsilon' < \kappa_0.(\xi+1)$, then as in the proof of 3.13, we see that $I_{\varepsilon'}(T)$ is α-finite. By 3.18, the requirements (R and S) of priority ε' and those of higher priority reach their final positions before stage $\beta_{\xi+1}$. It follows that the sequence $\langle I_\varepsilon(T) | \varepsilon < \varepsilon'\rangle$ is a simultaneously α-r.e. sequence of α-finite sets with parameters in $\beta_\xi+1$. By α-stability, this is in fact a $\beta_{\xi+1}$-finite sequence of $\beta_{\xi+1}$-finite sets. Now suppose $\varepsilon' < \kappa_0.(\xi+1)$ and for every $\varepsilon < \varepsilon'$, $\bigcup I_\delta(T)$ $(\delta < \varepsilon)$ is $\beta_{\xi+1}$-finite. Since κ_0 is the greatest α-cardinal, and also since by 3.18 requirements of higher priority than ε' reach their final positions before stage $\beta_{\xi+1}$, the sequence $\langle I_\delta(T) | \delta < \varepsilon'\rangle$ is (as in 3.14) α-finite (after these requirements have reached their final positions in priority, and after stage β_ξ, each of them will be injured, if at all, less than κ_0 many times, and hence less than a regular α-cardinal number of times). By α-stability, this sequence is $\beta_{\xi+1}$-finite. We conclude that for every $\varepsilon < \kappa_0.(\xi+1)$, $\bigcup I_\delta(T)$ with $\delta < \varepsilon$ is a subset of $\beta_{\xi+1}$. This implies the lemma.

It follows from Lemma 3.19 that requirement of priority higher than $\kappa_0.\xi$ does not get injured after stage β_ξ. The analogs of Lemmas 3.16 and 3.17 may now be proved similarly. This gives Theorem 3.5 for the case $\alpha^* = \alpha$ such that α has a greatest cardinal and the set of α-stable ordinals has order type less than α.

We now consider the second possibility in Case 2. Here we

have $\alpha^* = \alpha$, κ_0 is the greatest α-cardinal, and the set of stable ordinals has order type α. We are almost back to the case when α is a limit of α-cardinals. The priority system is now given by the identity function, so that T_d has higher priority than T'_e if $d < e$ or if $d = e$ and $T = R$, $T' = S$. Construction now proceeds as in the case when α is the limit of α-cardinals (which in turn is similar to Case 1). The fact that every $I_e(T)$ is α-finite is a consequence of the following:

<u>3.20 LEMMA</u> If $\alpha = \alpha^*$, κ_0 is the greatest cardinal of α, and the set of α-stable ordinals has order type α, then for all e, $I_d(T)$, $d < e$, is α-finite.

Proof. This is proved by induction. Passing from e to e+1 is almost identical to what has been done above. Observe however the following point: Beyond the least stage where $I_d(T)$, $d < e$, has stabilized (i.e. no longer grows), R_e may be injured at most e times. It follows that if $\bigcup I_d(T)$, $d < e$, is a subset of an α-stable ordinal, then $\bigcup I_d(T)$, $d < e$, is a subset of the next α-stable ordinal after e. From this and by induction hypothesis, one concludes that for each limit ordinal e, $\bigcup I_d(T)$, $d < e$, is a subset of the eth α-stable ordinal. Since the set of α-stable ordinals has order type α, we have the lemma.

Using 3.20 one may now show that the α-r.e. sets A and B have incomparable α-degrees. We leave the details to the reader. The proof of Theorem 3.5 is now complete.

<u>3.21 Remark</u> One may introduce more complications to the construction in order to make the sets A and B both hyperregular. See Sacks and Simpson [1972] for details.

<u>3.22 Remark</u> Lerman [1972] has shown, using an α-finite

injury argument, that every α-recursive partially ordered set is order isomorphic to a subset of the α-degrees of α-r.e. sets. As a consequence one obtains the corollary that a Σ_1 sentence in the theory of α-degrees of α-r.e. sets is true for one admissible ordinal if and only if it is true for all admissible ordinals. This is however false for Σ_2 sentences (Shore [1978]).

Section 2: Sacks Splitting

Sacks [1963a] proved that every Turing degree of a non-recursive r.e. set is the join of two other such degrees. In modern terminology, this says that every non-zero r.e. degree splits over 0. This result is obtained by 'splitting' a given non-recursive r.e. set C into the disjoint union of two r.e. sets A and B of lower Turing degree in neither of which is C recursive. While the method used in the proof is broadly speaking that of finite injury priority argument, it differs in a very essential way from those of the Friedberg-Muchnik type: There is no a priori recursive bound on the number of injuries inflicted on each requirement. In classical recursion theory this difference does not appear to lend any significance to the method of construction, since each requirement is still injured only finitely often, so that, by the regularity of ω, for each requirement there is a stage where it ceases to injure any requirement of lower priority. In higher recursion theory, however, the absence of a priori bounds necessitates the introduction of a new technique.

Let us recall the key points in the Sacks splitting construction. The objective is to obtain r.e. sets A and B so that $C = A \cup B$, $A \cap B = \emptyset$, and requirements $R_e : \{e\}^A \neq C$, and $S_e : \{e\}^B \neq C$ for all e. Give the requirements their natural priority with the proviso that for the same e, R_e has higher priority than S_e. Define A^n, B^n and C^n to be the respective sets enumerated at stage n. Let c^{n+1} be the element of C enumerated at stage n+1. The purpose is to put it in either A or B (this immediately implies that A and B are disjoint r.e., each recursive in C, with union equal to C). The basic strategy is to preserve as much as possible the length of agreement between $\{e\}^A$, say, and C. Thus if at stage n, we have

$$(3.23) \qquad \{e\}_n^{A^n}(x) = C^n(x),$$

for all $x < m$, where $m < n$ (we say in this case that $\{e\}^A$ and C agree on the initial segment m at stage n), we will try to protect this computation by insisting that no number specifically excluded from A in the computation of (3.23) may enter A at any stage after n with a view to satisfying a requirement of lower priority. Thus if the equality in (3.23) becomes an inequality for some x at a stage after n, it is due either to the attempt to satisfy a requirement of higher priority, or x is enumerated in C. We argue however that for no e can one have $\{e\}^A = C$. This is achieved by induction on e. Suppose this is true for all requirement of higher priority. Go to a stage n where for all $d < e$, there is an $x(d)$ where for $X = A$ or B, either

$$(3.24) \qquad \{d\}_s^{X^s}(x(d)) \text{ is undefined for all } s > n,$$

or

$$(3.25) \qquad \{d\}^X(x(d)) = \{d\}_s^{X^s}(x(d)) \neq C(x(d)) \quad \text{for all } s > n.$$

(We say that R_d, or S_d as the case may be, is eventually permanently satisfied). Suppose that n is chosen to be greater than the elements used in the computations of (3.24) and (3.25). Clearly then after stage n, any computation used to obtain an agreement between $\{e\}^A(x)$ and C(x) will be preserved, as this requirement has the highest priority. If $\{e\}^A = C$, every computation of the form $\{e\}^A(x)$ will be preserved, and this allows one to argue that C is in fact α-recursive, contradicting our assumption on C. In this way one shows that every requirement is eventually permanently satisfied.

Observe now the complexity of the function s: $e \to n(e)$, where n(e) is the least stage such that R_e and S_e are permanently satisfied. It is a Σ_2 function and there is no effective bound on the value of n(e). Consider the situation in α-recursion. If α is not Σ_2 admissible, there need not be a uniform upper bound for $\{\sigma(d) | d < e\}$, where $\sigma(d)$, the analogue of n(d), is the least stage such that requirements R_d and S_d are permanently satisfied. This prevents one from arguing by induction that $\sigma(d)$ exists for every $d < \alpha$. A uniform bound, however, does exist if one could somehow introduce a system of priority at most the length of $\Sigma_2(L_\alpha)$ cofinality. For then the $\Sigma_2(L_\alpha)$ function will ensure that sup $\{\sigma(d)\}$, $d < e$, is less than α for every e $< \sigma 2cf(\alpha)$, where now d represents requirements of priority d. However since there are α-many requirements of the types R_e and S_e, it is not altogether clear how these may be given a priority system of length $\sigma 2cf(\alpha)$, as in general $\sigma 2cf(\alpha) < \alpha$, and the two ordinals may have different cardinalities.

Shore [1975a] devised a way out of this difficulty. The idea is to subdivide the set of requirements into $\sigma 2cf(\alpha)$ many blocks. Those belonging to the same block are, so to speak, given the same priority. At each step of the construction, elements are put into either A or B depending on which blocks of requirements are to be protected (in order that the computations associated with those blocks be preserved). When this is faithfully carried out, one may argue that each requirement is eventually permanently satisfied in the following way. Let $blk(\zeta)$ denote the ζth block of requirements. Suppose that a bound $\sigma(blk(\zeta))$ exists for every block $\zeta < \xi$. Then as observed earlier, the function $\zeta \to (\mathrm{blk}(\zeta))$, being $\Sigma_2(L_\alpha)$, implies that a uniform bound on $\{\sigma(\mathrm{blk}(\zeta)) \mid \zeta < \xi\}$ also exists. One now argues that $\sigma(\mathrm{blk}(\xi))$ exists as well. This is argued by way of contradiction. If a bound does not exist, then one engages the help of the whole block $\mathrm{blk}(\xi)$ of requirements to effectively compute C, similar to the way one did for one requirement in the classical case. The fact that requirements in $\mathrm{blk}(\xi)$ are accorded equal priority ensures that a computation that is preserved for one requirement is not injured by a computation for another requirement in the same block. In this way the computations preserved for each of the requirements in the same block may be stacked up on one another to be used for the computation of the set C. One may then conclude that C is α-recursive, giving us the contradiction required.

This method, known as the 'blocking technique', essentially takes care of the problem of bounds in the construction. There is another problem which arises quite often in α-recursion. Let us return to the sketch above where we described how a bound on $\mathrm{blk}(\xi)$ was proved to exist. Somewhere in the proof, before arguing that C may be effectively computed, one has to single out those requirements in $\mathrm{blk}(\xi)$ which satisfy (3.25) (where n is now replaced by σ). These requirements give the wrong information about C and should therefore be excluded in the computation of C. It turns out that this set of requirements is in fact a bounded α-r.e. set. An obvious solution to the problem would then be to make this set α-finite. This will follow if it is bounded below α^*. We do this by indexing the set of requirements with a list of length α^*, while each block will be a bounded α-finite subset of α^*.

Finally there is the difficulty that C may not be regular. This will be required at some point in the proof. However, by Theorem 1.22 there is an α-r.e. α-regular set in the degree c of C, and so one may still conclude that **c** is the joint of two lesser α-r.e. degrees.

We now proceed with the proof of the Splitting Theorem.

3.26 THEOREM Let C be an α-r.e. regular non α-recursive set. Then C is the disjoint union of two α-r.e. sets A and B of lower α-degree than that of C.

Proof. By Lemma 2.7, there is a $\Sigma_2(L_\alpha)$ map from $\sigma 2\mathrm{cf}(\alpha)$ cofinally into α^*. Denote this map by f, and its α-recursive approximation by f'. Clearly it is possible to make $f'(\sigma,\zeta) \geq f'(\tau,\xi)$ whenever $\sigma \geq \tau$ and $\zeta \geq \xi$. We may therefore assume that f' has this property and that f is strictly increasing. We say that a requirement R_e or S_e belongs to block ζ at stage σ if we have $f'(\sigma,\zeta) \leq p_1(e) < f'(\sigma,\zeta+1)$. Let $blk(\sigma,\zeta)$ be the set of requirements that belong to block ζ at stage σ. Then for each ζ, $\lim_\sigma blk(\sigma,\zeta)$ exists and is equal to $\{p_1(e) \mid f(\zeta) \leq p_1(e) < f(\zeta+1)\}$. For future reference, recall also that for all $\rho < \alpha^*$, there is a stage σ where $p_1(e) > \rho$ for each $e \geq \sigma$.

We first set $A^{<\sigma} = \bigcup A^\tau$, $\tau < \sigma$. Define this similarly for B and C. The requirement T_e (T = R or S) is said to be satisfied at the beginning of stage σ if there is an ordinal $x(\sigma,e)$ such that for X = A (if T = R) or B (if T = S), either

$$(3.27) \qquad \{e\}_\sigma^{X^{<\sigma}}(x(\sigma,e)) \neq C^{<\sigma}(x(\sigma,e)) \text{ or}$$

$$(3.28) \qquad \{e\}_\sigma^{X^{<\sigma}}(x(\sigma,e)) \text{ is undefined.}$$

Our objective is to (intentionally) preserve the computations associated with requirements which are either <u>not</u> satisfied or meet condition (3.27).

Let c^σ be the element of C enumerated at stage σ. Find the least ordinal ζ (if exists) for which there is an e in $blk(\sigma,\zeta)$ such that either (3.27) holds for T_e (T = R or S) or T_e is not satisfied, and c^σ is an element specifically excluded in the computation (i.e. the fact that c^σ is not in $A^{<\sigma}$ (or $B^{<\sigma}$ as the case may be) is used in the computation). Put c^σ in B so that $B^\sigma = B^{<\sigma} \cup \{c^\sigma\}$ if the above holds for R_e. Put it in A otherwise (thus $A^\sigma = A^{<\sigma} \cup \{c^\sigma\}$).

If c^σ is put in A (B), let $B^\sigma = B^{<\sigma}$ ($A^\sigma = A^{<\sigma}$). If the ordinal ζ does not exist (so that c^σ does not interfere with any computation to be preserved), put c^σ in A, and define A^σ, B^σ accordingly. This completes the construction at stage σ.

$R_e(S_e)$ is said to be satisfied from stage σ_0 on if there is an x(e) such that (3.27) or (3.28) holds for all $\sigma > \sigma_0$ when $x(\sigma,e)$ is set to be x(e). We prove by induction on blocks $\mathrm{blk}(\zeta)$ that each requirement is eventually permanently satisfied.

<u>3.29 LEMMA</u> Let $\lambda < \sigma2fc(\alpha)$ be a limit ordinal. Suppose that for each $\zeta < \lambda$, there is a stage $\sigma(\zeta)$ where each requirement in $\mathrm{blk}(\zeta)$ is satisfied from $\sigma(\zeta)$ on (with the same $x(\sigma,e)$). Then $\sup\{\sigma(\zeta) \mid \zeta < \lambda\}$ is bounded below α.

Proof. Observe that (3.27) and (3.28) imply that the function $\zeta \to \sigma(\zeta)$ is $\Sigma_2(L_\alpha)$. Since $\lambda < \sigma2cf(\alpha)$, the supremum of these ordinals is clearly less than α.

<u>3.30 LEMMA</u> Fix $\xi < \sigma2fc(\alpha)$. Let σ_o be a stage where every requirement in $\mathrm{blk}(\zeta)$, $\zeta < \xi$, is satisfied from σ_o on (with the same $x(\sigma_o,e)$). Then there is a stage $\sigma_1 > \sigma_o$ such that every requirement in $\mathrm{blk}(\xi)$ of type R is satisfied from σ_1 on.

Proof. It will be shown that if such a bound does not exist, then C is α-recursive.

First of all note that there is a stage σ where $f'(\sigma,\zeta) = f(\zeta)$ for all $\zeta < \xi$. There is also a stage τ where the set $(p_1^{-1})"f(\xi+1)$ is contained in τ. We may assume for convenience that at stage σ_o both of these have occurred, so that $blk(\sigma_o,\xi) = blk(\xi)$. We may also assume that all the ordinals involved in the computations associated at stage σ_o with requirements in $blk(\zeta)$, $\zeta < \xi$, are less than σ_o.

Let K be the set of all ε's such that $p_1^{-1}(\varepsilon)$ lies in $blk(\xi)$ and (3.27) holds for R_e ($e = p_1^{-1}(\varepsilon)$) at some stage $\sigma > \sigma_0$ with $C^{<\sigma}(x(\sigma,e)) = 1$ (meaning $x(\sigma,e) \in C$). K is a bounded α-r.e. subset of α^* and so is α-finite. For each $p_1(e)$ in K, let $\sigma(e)$ be the least stage after σ_0 where $p_1(e)$ is enumerated in K. This is an α-recursive map, and so the range is α-finite bounded by a stage σ', say. Since each $\sigma(e)$, for $e \in K$, is greater than σ_o, by the construction the computations associated with R_e at stage $\sigma(e)$ will be preserved forever. (If a computation is destroyed, it will be those associated with requirements of type S in $blk(\xi)$). In particular after stage σ', the inequality displayed in (3.27) for R_e will stay. Notice moreover that by the α-regularity of C, the set $C|\sigma'$ is α-finite.

Now for every $x > \sigma'$ and every stage $\sigma > \sigma'$, if e belongs to $blk(\xi) - K$, and if $\{e\}_\sigma^{A^{<\sigma}}(x)$ is both defined and agrees with $C^{<\sigma}(x)$, then this computation is henceforth preserved, implying that the agreement will stay at all subsequent stages. Suppose now a bound σ_1 as required in the lemma does not exist. This implies that for every $x > \sigma'$, there is an e in $blk(\xi) - K$ and a stage $\sigma > \sigma'$ such that $\{e\}_\sigma^{A^{<\sigma}}(x)$ is defined and agrees with $C^{<\sigma}(x)$. We give an algorithm to compute C beyond σ' in an α-recursive way. To obtain $C|\nu$, where $\nu > \sigma'$, go to a stage $\tau(\nu)$ such that for every x between σ' and ν, there is an e in $blk(\xi) - K$ for which $\{e\}_{\tau(\nu)}^{A^{\tau(\nu)}}(x)$ is defined and agrees with $C^{\tau(\nu)}(x)$. Such a stage $\tau(\nu)$ exists by the admissibility of α. By the choice of the stage σ', this agreement will remain so thereafter. This implies that it actually gives the correct information about C(x). Together with the fact that $C|\sigma'$ is α-finite, this shows that C is α-

recursive, a contradiction. Thus there is a stage σ_1 where every requirement of type R in blk(ξ) is satisfied thereafter. This proves Lemma 3.30.

3.31 LEMMA Under the same assumption as Lemma 3.30, there is a stage σ_2 where every requirement of type S in blk(ξ) is satisfied from stage σ_2 onwards.

Proof. Let σ_0 and σ_1 be as given in Lemma 3.30. By the α-regularity of C, there is a stage where $C|\sigma_1$ is enumerated. Without loss of generality, let us assume that σ_1 has this property and that it is greater than all the ordinals used in the computations associated with requirements of type R and S in blk(ζ), $\zeta < \xi$, and requirements of type R in blk(ξ). Then at no stage after σ_1 will there be an element enumerated in C which is put in B for the sake of preserving computations associated with these requirements. By our assumption, after stage σ_1, every computation associated with a requirement of type S in blk(ξ) will be preserved. The proof now proceeds in a way similar to that given in Lemma 3.30, whereby one argues that if the stage σ_2 does not exist, then C is in fact α-recursive.

We may now complete the proof of Theorem 3.25.

3.32 LEMMA $C \not\leq_{w\alpha} A$ and $C \not\leq_{w\alpha} B$.

Proof. We show that $C \not\leq_{w\alpha} A$. Suppose that $\{e\}^A = C$. Let e be in blk(ξ). By Lemma 3.30 there is a stage σ_1 where R_e is satisfied from σ_1 onwards. This means that either (3.27) or (3.28) holds for the same $x(\sigma_1, e)$ at all stages $\sigma > \sigma_1$. So either there is an ordinal x where $\{e\}^A(x)$ does not compute C(x) correctly, or the function $\{e\}^A$ is not total. The conclusion follows.

3.33 LEMMA $C = A \cup B$ and $A \cap B = \emptyset$.

Proof. This follows from the construction.

We now derive some corollaries.

3.34 COROLLARY Every α-r.e., non α-recursive degree **c** is the joint of two incomparable α-r.e. degrees.

Proof. By Theorem 1.22 one may choose a representative C in **c** which is α-regular. By Theorem 3.26, C is the disjoint union of two α-r.e. sets of strictly lower degrees a and **b**. Now a and **b** are not comparable since otherwise their joint, which is equal to **c**, will again be equal to one of them.

3.35 COROLLARY No α-r.e.degree is minimal, i.e. below every non-zero α-r.e. degree there is a non-zero α-r.e. degree.

Proof. This follows from Corollary 3.34.

3.36 Remark By exploiting the idea of tame $\Sigma_2(L_\alpha)$ projectum introduced by Lerman, Chong [1976] independently gave a different proof of Corollary 3.35.

CHAPTER 4

MAXIMAL SETS

The class of α-r.e. sets form a lattice under the operations of union and intersection. For $\alpha = \omega$, this class has been extensively investigated over the past thirty years. Among the sets which have received particular attention are simple sets, hyperhypersimple sets, major subsets and maximal sets. In particular the notion of a maximal set has played a major role in classicl resursion theory (Soare [1975]). It is not surprising then that these are the sets which were first studied in the context of the lattice of α-r.e. sets for arbitrary α.

In this chapter we prove Lerman's result [1974] which gives a necessary and sufficient condition (on α) for the existence of a maximal set. The earliest discovery that not every admissible ordinal has a maximal set was made in [Sacks 1966a]. It indicated for the first time that not every construction introduced in classical recursion theory could be lifted to higher ordinals. The complete characterization given by Lerman showed that indeed the existence of maximal sets is very closely tied to the notion of countability. This result is important in two ways: Firstly, it enhances our understanding of the crucial role played by the ordinal ω in some fundamental constructions in recursion theory. It is a vindication of Kreisel's expressed belief that one of the reasons for generalizing recursion theory is to improve our understanding of recursion theory on ω ([Kreisel 1971]). Secondly, the absence of maximal sets in some admissible ordinals indicates an important structural difference in the lattice of α-r.e. sets for various α. This non-homogeneity behaviour compels one to view recursion theory on admissible ordinals to be more than just a generalization of results from classical recursion theory using fine structure theory.

An α-r.e. set W is <u>simple</u> if α - W is not α-finite and every α-r.e. subset of α - W is α-finite. This notion was introduced by Post in his attempt to construct a non-recursive, incomplete r.e. set. It is not difficult to construct a simple set for every α. Indeed one can show that the 'deficiency set'.

$$D = \{x \mid (\exists\nu)[\nu > x \ \& \ p_1(\nu) < p_1(x)]$$

is simple and that α - D has order type less than or equal to α^*. A maximal set is a simple set whose complement is in a sense 'sparse'. This is made precise by the following definitions.

<u>4.1 DEFINITION</u> An α-finite set is α^*-finite if it has order type less than α^*.

Equivalently, and this is the motivation for the definition, K is α^*-finite if and only if every α-r.e. subset of K is α-recursive.

<u>4.2 DEFINITION</u> An α-r.e. set M is maximal if α - M is not α^*-finite and if for every α-r.e. set $W \supseteq M$, either α - W or W - M is α^*-finite.

It should be noted that there are several possible definitions for maximal α-r.e. sets. This is discussed at length in Lerman [1974]. The reason that α^*-finite sets are chosen over α-finite sets is lattice-theoretic. Specifically, the most significant uses of α-r.e. maximal sets have been in the study of the lattice of r.e. sets (for example, in Lachlan [1968]). The class of α^*-finite sets is definable in the language of the lattice of α-r.e. sets, and the α-r.e. maximal sets are precisely the maximal elements in the lattice of α-r.e. sets modulo α^*-finite sets. Lerman [1974] examined various possible definitions for maximal sets, and proved necessary and sufficient conditions for their existence. We will only consider maximal sets defined in the sense above.

<u>4.3 PROPOSITION</u> Let M be maximal. Then M is not α-recursive, so that α - M is not α-finite.

Proof. Suppose that M is α-recursive. By definition α - M is not α^*-finite and so is of order type at least α^*. Since either $\alpha^* = \alpha$ or is an α-cardinal (Lemma 2.8), one may split α - M into two disjoint α-recursive sets each of order type at least α^*. Call these sets U_1 and U_2. Let $W = M \cup U_1$. Then W violates the maximality of M in Definition 4.2.

<u>4.4 PROPOSITION</u> Let M be maximal. Then M is simple.

Proof. Let U be an α-r.e. subset of α - M. Since M is maximal, either U is α^*-finite (hence α-finite) or $\alpha - (M \cup U)$ is α^*-finite. The latter implies that M is α-recursive, contradicting Proposition 4.3.

Recall the definition of $S_3(L_\alpha)$ projectum (1.52). Lerman's characterization of the existence of maximal sets is given in terms of this ordinal.

<u>4.5 THEOREM</u> There is a maximal set in α if and only if $\omega = S_3(L_\alpha)$ projectum.

The construction of a maximal set uses the technique of 'e-state' invented by Friedberg [1958] for ω-recursion theory. In higher recursion theory, Kreisel and Sacks [1965] first constructed maximal sets for the ordinal ω_1^{CK}. Lerman and Simpson [1973] extended this result to every α such that $\sigma 2p(\alpha) = \omega$. In each case the method used was essentially the e-state method. By the proof of Theorem 4.5, this method is roughly speaking the only method that is needed for the construction of maximal sets. In all cases of α where the method fails to work, there is no maximal α-r.e. set. Methodologically speaking this shows both the power and limitations of the e-state method, and the surprising link of the notion of maximal sets to that of S_3 countability. This is a significant aspect of α-recursion theory revealed in Theorem 4.5.

Since the proof of the theorem is fairly long, we shall divide it into two parts, devoted respectively to existence and non-existence results.

Section 1: Existence of Maximal Sets

For the rest of the chapter, we fix λ to be the Σ_2 cofinality of L_α, and assume that $S_3(L_\alpha)$ projectum $= \omega$.

We first prove a lemma due to Simpson (Simpson [Notes]).

4.6 LEMMA There is a $\Sigma_2(L_\alpha)$ function $g\colon \lambda \to \omega$ such that for all n, the set $\{\nu | \nu < \lambda \;\&\; g(\nu) \leq n\}$ is bounded below λ.

Proof. We assume that $\omega < \lambda$, since otherwise the lemma is immediate. Let $f\colon \lambda \to \alpha$ be a $\Sigma_2(L_\alpha)$ cofinality function. By Theorem 1.55 let $j\colon \omega \to \alpha$ be an $S_3(L_\alpha)$ surjection approximated by the S_2 function $j'(\sigma,n)$. Thus $j(n) = \lim_\sigma j'(\sigma,n)$. By Theorem 1.50 j' is $\Sigma_2(L_\alpha)$.

Claim For all $\nu < \lambda$, there is an η, $\nu \leq \eta < \lambda$, such that for some n,

$$(4.7)\quad j'(f(\eta),n) \neq \lim_{\zeta\to\eta} j'(f(\zeta),n),$$

if the limit on the right hand side exists.

Suppose that the claim is false. Then there is a

ν_0 such that for all η, $\nu_0 \leq \eta < \lambda$, for all n, if $\lim_{\zeta\to\eta} j'(f(\zeta),n)$ exists, it is equal to $j'(f(\eta),n)$. By the definition of j, for each n there exists an $\eta > \nu_0$ such that $\lim_{\zeta\to\eta} j'(f(\zeta),n)$ exists. It follows that the first time when such limit occurs it gives the correct value of j(n). In particualr, since for η a successor ordinal, $j'(f(\eta),n)$ is equal to $\lim_{\zeta\to\eta} j'(f(\zeta),n)$. One concludes that $j(n) = j'(f(\nu_0),n)$ for all n. Since j'is $\Sigma_2(L_\alpha)$, this gives us a $\Sigma_2(L_\alpha)$ cofinal function (in fact a surjection) from ω onto α, contradicting the assumption that $\omega < \lambda$. This proves the claim.

We may now define the function g. For each $\nu < \lambda$, let $\eta(\nu)$ be the least ordinal η, $\nu \leq \eta < \lambda$, such that (4.7) holds for some n. Let $g(\nu)$ be the least n such that (4.7) holds for $\eta = \eta(\nu)$. The reader may verify that g is $\Sigma_2(L_\alpha)$ (the fact that g is defined on λ is used in an essential way).

To see that g satisfies the conclusion of the lemma, let n be fixed. By the definition of j, there is a $\nu(n)$ such that $j(n) = j'(f(\nu),n)$ for all ν such that $\nu > \nu(n)$. Then $g(\nu) \neq n$ for all such ν. This implies that the set $\{\nu | \nu < \lambda \ \&\ g(\nu) \leq n\}$ is bounded below λ

<u>4.8 THEOREM</u> If the S_3 projectum of α is ω, then there is a maximal set.

Proof We follow the exposition given in Simpson [Notes]. Let λ, f, g, j be as given in Lemma 4.6. Let f', g' be the α-recursive approximations of f and g respectively, and let j' be the $S_2(L_\alpha)$ approximation of j ($\lim_\sigma j'(\sigma,n) = j(n)$) and let j" be α-recursive so that $\lim_{\tau\to\alpha} j''(\sigma,\tau,n) = j'(\sigma,n)$. We shall construct an α-r.e. set M such that: (*) $\alpha - M$ is unbounded in α and has order type λ, and given an α-r.e. set $W \supseteq M$, either $W - M$ is α-finite or $\alpha - W$ is α-finite. This set M will be maximal since for any $W \supseteq M$, $W - M$ is α-finite if and only if it is α-finite and of order type less than λ (hence less than α^*). The same holds for $\alpha - W$.

At stage σ, let $m'(\sigma,\nu)$, denote the potential value of the νth element (in ascending order) of $\alpha - M$ (one may also think of this as the position of the marker for ν at the end of stage σ). The function $m'(\sigma,\nu)$ has the property that $m'(\sigma,\nu) \geq m'(\tau,\eta)$ for $\sigma \geq \tau$ and $\nu \geq \eta$. We ensure that $m(\nu) = \lim_\sigma m'(\sigma,\nu)$ exists for each ν and we make $M = \alpha - \{m(\nu) | \nu < \lambda\}$.

Let $g'(\sigma,\nu) = e$. We define the $g(\nu)$-state function $E: \alpha\times\lambda\times\alpha \to \omega$ by

$$E(\sigma,\nu,x) = \Sigma\{2^{e-i} | i \leq e \ \& \ x \in W^{\sigma}_{j''(f(\sigma,\nu),\sigma,i)}\}.$$

Observe that the function E is α-recursive. Furthermore by our convention (introduced in Chapter 1), all the ordinals enumerated in an α-r.e. set at the stage σ are less than σ. It follows that $E(\sigma,\nu,x)$ is not defined for $x > \sigma$.

Set $M^{<\sigma} = \bigcup M^{\tau}$, $\tau < \sigma$. Inductively define, for each ν, the largest integer k, denoted $k'(\sigma,\nu)$, such that for some $x \notin M^{<\sigma}$,

$$(4.9) \qquad E(\sigma,\nu,x) = k$$

$$(4.10) \qquad x > m(\sigma,\zeta) \text{ and } x > f'(\sigma,\zeta) \text{ for all } \zeta < \nu.$$

Let $m'(\sigma,\nu)$ be the least ordinal x satisfying (4.9) and (4.10) for $k = k'(\sigma,\nu)$ (hence $E(\sigma,\nu,m'(\sigma,\nu)) = k(\sigma,\nu)$). Let $z(\sigma)$ be $\max \{m'(\sigma,\zeta)\} + 1$, where the maximum is taken over all $\zeta < \nu$ for which $m'(\sigma,\zeta)$ is defined (if $\sigma < \lambda$ then not every $m'(\sigma,\zeta)$, $\zeta < \lambda$, is defined at stage σ). Let $M^{\sigma} = z(\sigma) - \{m'(\sigma,\nu) | m'(\sigma,\nu) \text{ is defined}\}$. Then $M^{\sigma} \supseteq M^{<\sigma}$. Notice that $m'(\sigma(\nu)$, if defined, is the νth member of

$\alpha - M^{\sigma}$ (in increasing order).

This completes the construction at stage σ. Set $M = \bigcup M^{\sigma}$, $\sigma < \alpha$.

The next few lemmas will verify that M satisfies property (*) and hence maximal.

<u>4.11 LEMMA</u> For each $\nu < \lambda$, $\lim_{\sigma} m'(\sigma,\nu) = m(\nu)$ exists.

Proof. We prove this by induction. Suppose that the limit exists for each $\zeta < \nu$. Since $\lambda = \sigma 2cf(\alpha)$, $\sup\{m(\zeta)|\zeta < \nu\}$ exists and is less than α. In particular there is a stage σ_0 where $m'(\sigma,\zeta) = m(\zeta)$ for all $\zeta < \nu$ and $\sigma \geq \sigma_0$. Assume also that σ_0 has been chosen to be a stage where $f'(\sigma,\zeta) = f(\zeta)$ and $g'(\sigma,\zeta) = g(\zeta)$ for all $\zeta \leq \nu$ and $\sigma \geq \sigma_0$ (recall that f and g are $\Sigma_2(L_\alpha)$ functions defined on λ). Then for $\sigma > \tau \geq \sigma_0$, the value of $m'(\sigma,\nu)$ differs from $m(\tau,\nu)$ only when $E(\sigma,\nu,m'(\sigma,\nu)) > E(\tau,\nu,m'(\tau,\nu))$ (since the value is no longer affected by the values of $m'(\sigma_0,\zeta)$, $f'(\sigma_0,\zeta)$, and $g'(\sigma_0,\zeta)$ for any $\zeta < \nu$, so that (4.10) is automatically satisfied). But notice that $E(\sigma,\nu,m'(\sigma,\nu))$ is an integer bounded above by $\Sigma\, 2^{e-i}$, $i \leq e$. This implies that indeed there is a stage $\sigma_1 \geq \sigma_0$ such that $m'(\sigma,\nu) = m'(\sigma_1,\nu)$ for all $\sigma \geq \sigma_1$.

It follows from Lemma 4.11 that for each $\nu < \lambda$, $\lim_{\sigma} k'(\sigma,\nu) = k(\nu)$ exists. This in turn implies that $\lim_{\sigma} E(\sigma,\nu,m'(\sigma,\nu)) = E^*(\nu,m'(\nu))$ exists as well and is in fact equal to $k(\nu)$. This follows from the fact that

$$\lim_{\sigma} E(\sigma,\nu,m'(\sigma,\nu)) = \Sigma\{2^{e-i}|i \leq e \;\&\; m(\nu) \in W_{j'(f(\nu),i)}\}.$$

<u>4.12 LEMMA</u> If $\nu < \eta < \lambda$, then $E^*(\nu,m(\nu)) \geq E^*(\eta,m(\eta))$.

Proof. Let σ_0 be a stage such that $m'(\sigma,\nu) = m(\nu)$, $m'(\sigma,\eta) =$

$m(\eta)$, and $E(\sigma,\nu,m(\nu)) = E^*(\nu,m(\nu))$, $E(\sigma,\eta,m(\eta)) = E^*(\eta,m(\eta)) = k$ for all $\sigma \geq \sigma_0$. By construction, $m'(\sigma,\nu) > m'(\sigma,\eta)$. Let $x = m(\eta)$. Then (4.10) holds and so by stage σ_0 at least one would set $k'(\sigma_0,\nu) = k$. Since the construction aims to 'maximize' the $g(\nu)$-state of $m(\nu)$, by our choice of the stage σ_o we have $k'(\sigma,\nu) \geq k'(\sigma_0,\nu)$ for all $\sigma > \sigma_0$. Hence $E^*(\nu,m(\nu)) \geq E^*(\eta,m(\eta))$.

The next lemma imposes strong restrictions on the set $\alpha - M$.

<u>4.13 LEMMA</u> For each $i < \omega$, there is a $\nu(i) < \lambda$ such that if $\{m(\nu)|\nu\geq\nu(i)\} \cap W_{j(i)}$ is unbounded in α then

$$\{m(\nu)|\nu \geq \nu(i)\} \subseteq W_{j(i)}.$$

Proof. Assume that $\nu(i)$ exists for each $i < i_0$. Let K be the (finite) set of integers $i < i_o$ such that $\{m(\nu)|\nu \geq \nu(i)\} \cap W_{j(i)}$ is bounded in α by $\sigma(i)$ ($\geq \nu(i)$). Then for $i < i_o$ and $i \notin K$, we have $\{m(\nu)|\nu \geq \nu(i)\} \subseteq W_{j(i)}$. Let

$$\nu^* = \max\ \{\sup\ \{\sigma(i)|i \in K\},\ \sup\ \{\nu(i)|i < i_0\ \&\ i \notin K\}\}.$$

Clearly $\nu^* < \lambda$. Let $\nu(i_o)$ be the least ordinal greater than ν^* such that

$$(4.14) \qquad g(\zeta) > i_0 \text{ for all } \zeta > \nu \text{ (Lemma 4.6)},$$

$$(4.15) \qquad j'(f(\zeta),i) = j(i) \text{ for } i < i_0 \text{ and } \zeta > \nu.$$

We claim that $\nu(i_0)$ has the desired property. Suppose that $\{m(\nu)|\nu > \nu(i_0)\} \cap W_{j(i)}$ is unbounded in α. Let $\zeta > \nu > \nu(i_0)$. Let $g(\nu) = e$.

By (4.14) $e \geq i_0$. By Lemma 4.12 the inequality $E^*(\nu,m(\nu)) \geq E^*(\zeta,m(\zeta))$ holds. Unravelling the definition of E^* we have

$$\Sigma\{2^{e-i} | i < e \ \& \ m(\nu) \ \varepsilon \ W_{j'(f(\nu),i)}\} >$$

$$\Sigma\{2^{e-i} | i < e \ \& \ m(\zeta) \ \varepsilon \ W_{j'(f(\nu),i)}\}.$$

By the fact that $\zeta > \nu > \nu^*$ and by (4.14) and (4.15) we have

$$\Sigma\{2^{e-i} | i_0 < i < e \ \& \ m(\nu) \ \varepsilon \ W_{j'(f(\nu),i)}\} >$$

$$\Sigma\{2^{e-i} | i_0 < i < e \ \& \ m(\zeta) \ \varepsilon \ W_{j'(f(\nu),i)}\}.$$

Now $2^{e-i_0} > \Sigma\{2^{e-i} | i_0 < i < e\}$. Hence if $m(\zeta) \ \varepsilon \ W_{j'(f(\nu),i_0)}$ then $m(\nu) \ \varepsilon \ W_{j'(f(\nu),i_0)}$. By (4.15) this implies that if $m(\zeta)$ lies in $W_{j(i_0)}$ then $m(\nu)$ is in $W_{j(i_0)}$. It follows that if $\{m(\nu) | \nu > \nu(i_0)\} \cap W_{j(i)}$ is unbounded in α, then $\{m(\nu) | \nu > \nu(i_0)\}$ is a subset of $W_{j(i_0)}$.

<u>4.16 LEMMA</u> M is a maximal set.

Proof. First of all $\alpha - M$ is unbounded and of order type λ. Let $W_{j(i)} \supset M$. If $W_{j(i)} - M$ is bounded in α, then it contains less than order type λ many elements of $\alpha - M$. This set is α-finite by construction and in fact since $\lambda < \alpha^*$ it is α^*-finite. On the other hand, if $W_{j(i)} - M$ is unbounded then by Lemma 4.13

$$W_{j(i)} \supset \{m(\nu) | \nu > \nu\ (i)\}$$

and so again only α^*-finitely many elements of $\alpha - M$ are omitted by $W_{j(i)}$.

Section 2: Non-existence of Maximal Sets

In this section we prove the converse of Theorem 4.8. This is split into two cases: whether the complement of a maximal set is bounded or unbounded. The former is due to Lerman ([1974]), and the latter to Leggett ([1974]).

4.17 THEOREM Suppose that M is a maximal α-r.e. set such that $\alpha - M$ is unbounded in α. Then $S_3(L_\alpha)$ projectum = ω.

We continue to let $\lambda = \sigma 2cf(\alpha)$.

4.18 LEMMA There is an α-recursive function $g\colon \alpha \to \lambda$ such that for all $\nu < \lambda$, the set $\{\sigma | g(\sigma) < \nu\}$ is bounded in α.

Proof. This is similar to, but simpler than, the proof of Lemma 4.6. Assume that $\lambda < \alpha$ since otherwise the lemma is immediate. Let $f\colon \lambda \to \alpha$ be an $S_2(L_\alpha)$ cofinal function approximated by the α-recursive function $f'\colon \alpha \times \lambda \to \alpha$.

Claim For all $\sigma < \alpha$, there exist $\tau > \sigma$ and $\nu < \lambda$ such that $\lim_{\zeta\to\sigma} f'(\zeta,\nu)$ exists and is not equal to $f'(\sigma,\nu)$.

Clearly for all σ and for all ν there is a $\tau > \sigma$ such that $\lim_{\zeta\to\tau} f'(\zeta,\nu)$ exists. If the claim is false, there is a σ_0 such that for all $\tau > \sigma_0$ and ν, if $\lim_{\zeta\to\tau} f'(\zeta,\nu)$ exists then it is equal to $f'(\tau,\nu)$. This implies that $f(\nu) = \lim_{\zeta\to\tau} f'(\zeta,\nu)$ for each ν, where $\tau > \sigma_0$ is an ordinal for which $\lim_{\zeta\to\tau} f'(\zeta,\nu)$ exists. But then f is α-recursive. By admissibility this is not possible since $\lambda < \alpha$.

Now for each σ let $\tau(\sigma)$ be the least $\tau < \alpha$ such that for some $\nu < \lambda$, $\lim_{\zeta\to\tau} f'(\zeta,\nu)$ exists and is not equal to $f'(\tau,\nu)$. Let $g(\sigma)$ be the least $\nu < \lambda$ such that $\lim_{\zeta\to\tau(\sigma)} f'(\zeta,\nu)$ exists and is not equal to $f'(\tau(\sigma),\nu)$. Clearly g is α-recursive. We show that g has the desired property. Let $\nu < \lambda$ be fixed. There is a stage σ_0 such that for all $\tau > \sigma_0$, $f'(\tau,\mu) = f(\mu)$ for all $\mu < \nu$. For $\tau > \sigma_0$, we get $g(\tau) > \nu$.

4.19 LEMMA If there exists an unbounded maximal set, then

there is an $S_2(L_\alpha)$ function from λ onto α.

Proof. Let g be as given in Lemma 4.18. Define

$$Y = \{\gamma | \gamma \in \alpha - M \ \& \ (\forall\delta>\gamma)[\delta \in (\alpha - M) \rightarrow g(\delta) > g(\gamma)]\}.$$

Observe that Y is unbounded in α. For suppose that Y is bounded. Inductively let δ_{n+1} be an ordinal in $\alpha - M$ and greater than δ_n such that $g(\delta_{n+1}) < g(\delta_n)$. This gives us an infinite descending sequence of ordinals, a contradiction.

Let $Y^* = g''Y$. Then $Y^* \subseteq \lambda$ and is $\Sigma_2(L_\alpha)$(notice that Y is $\Pi_1(L_\alpha)$). By the choice of g and the fact that Y is unbounded we see that Y^* is unbounded in λ.

Claim Y^* is not α-finite.

Otherwise, split Y^* into the disjoint union of two α-finite sets Y_1^* and Y_2^* each unbounded in λ. Let

$$Y_i = \{\gamma | g(\sigma) \in Y_i^*\}$$

for i = 1, 2. Then each Y_i is α-recursive and unbounded in α (Y_1 and Y_2 are also disjoint). Furthermore $M \cup Y_i$ contains unboundedly many elements of $\alpha - M$. Thus $(\alpha - M) \cap (M \cup Y_i)$ and $\alpha - (M \cup Y_i)$ are unbounded. This contradicts the maximality of M and so Y^* is not α-finite.

It follows from the claim that $\sigma 2p(\alpha) < \lambda$. By Claim 2 of Theorem 1.55, we have $\sigma 2cf(\alpha) = \delta 2p(\alpha)$ (since if $\sigma 2cf(\alpha) < \delta 2p(\alpha)$ then one has by Claim 2 of Theorem 1.55 that $\sigma 2p(\alpha) = \delta 2p(\alpha)$. This is not possible since $\sigma 2p(\alpha) < \sigma 2cf(\alpha) < \delta 2p(\alpha)$). By Corollary 1.53, $S_2(L_\alpha)$ projectum = $\delta 2p(\alpha)$. By Theorem 1.55 there is an $S_2(L_\alpha)$ function from λ onto α.

<u>4.20 LEMMA</u> Let M be α-r.e. such that $M = \bigcup M^\sigma, \sigma < \alpha$. For each $\gamma < \alpha$, there is a $\sigma(\gamma)$ such that

$$\text{order type } (\gamma - M) = \text{order type } (\gamma - M^\sigma)$$

for all $\sigma \geq \sigma(\alpha)$.

Proof. We prove this by induction. Let ν be the order type of $\gamma - M$. Clearly the lemma is true if ν is a successor ordinal. Suppose that ν is a limit ordinal and $\sigma(\delta)$ exists for all $\delta < \gamma$. Then for all $\delta < \gamma$, there is a stage σ where order type of the set $\delta - M^\sigma$ is less than ν (notice that order type $(\gamma - M) >$ order type $(\delta - M)$ for $\delta < \gamma$). This inequality continues to hold for all $\tau \geq \sigma$. Let $h(\delta)$ be the least stage σ where order type $(\delta - M^\sigma) < \nu$. Then h is α-recursive and so $h''\gamma$ is bounded in α. Let $\sup h''\gamma = \sigma(\gamma)$. Then we have order type $(\gamma - M) =$ order type $(\gamma - M^\sigma)$ for all $\sigma \geq \sigma(\gamma)$.

<u>4.21 LEMMA</u> If there is a maximal set M with unbounded complement, then λ is countable.

Proof. Let β be the order type of $\alpha - M$. Since $\alpha - M$ is unbounded and $\lambda = \sigma 2cf(\alpha)$, there is an ordinal $\xi < \alpha$ such that $\beta = \lambda.\xi$. We can split β into a simultaneously α-r.e. sequence of pairwise disjoint α-r.e. sets A_ν, $\nu < \lambda$, such that $\bigcup A_\nu$, $\nu < \lambda$, is equal to β, and each A_ν is unbounded in β. Let

$$B_\nu = \{\gamma | (\exists\delta)[\text{order type of } \gamma - M^\sigma] \in A_\nu\}.$$

Then B_ν is α-r.e. for each ν. Since each A_ν is unbounded in β, there exist unboundedly many ordinals γ in α such that the order type of $\gamma - M$ belongs to A_ν. By Lemma 4.20 for each such ordinal γ there is a σ such that the order type of $\gamma - M^\sigma$ is in A_ν. It follows that B_ν is an unbounded α-r.e. set containing unboundedly many elements of α -

M. By the maximality of M, $\alpha - (M \cup B_\nu)$ is α^*-finite. Let $h(\nu)$ be the sup of this α^*-finite set.

Claim for each γ, there exist only finitely many ordinals $\nu < \lambda$ such that $h(\nu) < \gamma$.

Suppose this is not the case. Then we can choose a γ in $\alpha - M$ such that $h(\nu) < \zeta$ for infinitely many $\nu < \lambda$. This means that $\gamma \in B_\nu$ for infinitely many ν. As the A_ν's are pairwise disjoint, there exist infinitely many stages σ such that the order types of $\gamma - M^\sigma$ are pairwise different. Now clearly order type $(\gamma - M^\sigma) \leq$ order type $(\gamma - M^\tau)$ for $\tau > \sigma$. It follows that there can be at most finitely many values assumed by the order type of $\gamma - M^\sigma$ for $\sigma < \alpha$ (otherwise one obtains an infinite descending sequence of ordinals which is impossible). This contradiction proves the claim.

The lemma is now a consequence of the claim, since if λ is uncountable, the conclusion of the claim must fail for at least one (hence all ordinals greater than) γ.

Lemmas 4.19 and 4.21 show that if there is a maximal α-r.e. set with unbounded complement, then λ, hence $S_2(L_\alpha)$ projectum, is countable. Our next task is to establish an $S_3(L_\alpha)$ bijection between λ and ω. Combining this with Lemma 4.19 we may conclude that $S_3(L_\alpha)$ projectum $= \omega$.

We define a linear ordering on λ. Given $\mu < \nu < \lambda$,

$$(4.22)\quad \mu <^* \nu \leftrightarrow h(\mu) < h(\nu) \text{ or } [h(\mu) = h(\nu) \;\&\; \mu < \nu],$$

where h is the function defined in Lemma 4.21. Then for each $\nu < \lambda$, there are only finitely many $\mu < \lambda$ such that $\mu <^* \nu$. Let $j(n)$ = the (n+1)st element of λ under the ordering $<^*$. Then $j: \omega \to \lambda$ is a bijection. We will show that j is $S_3(L_\alpha)$.

4.23 LEMMA The function j is an $S_3(L_\alpha)$ surjection from ω onto λ.

Proof. Recall first of all that $h(\nu) = \sup [\alpha - (M \cup B_\nu)]$. We define an α-recursive approximation of h in the following way: Set $h''(\sigma,\tau,\nu) = \sup [\sigma - (M^\tau \cup B_\nu^\tau]$. Set $h'(\sigma,\nu) = \lim_{\tau\to\alpha} h''(\sigma,\tau,\nu)$. Then $h'(\sigma,\nu) = \sup [\sigma - (M \cup B_\nu)]$. Observe now $\lim_{\sigma\to\alpha} h'(\sigma,\nu) = h(\nu)$ so that h is $S_3(L_\alpha)$. Next we define an approximation to the ordering $<^*$:

$$(4.24) \quad \mu <^*_{\sigma,\tau} \nu \leftrightarrow$$
$$h''(\sigma,\tau,\mu) < h''(\sigma,\tau,\nu) \text{ or}$$
$$[h''(\sigma,\tau,\mu) = h''(\sigma,\tau,\nu) \ \& \ \mu < \nu].$$

Set $j''(\sigma,\tau,n)$ = the (n+1)st element of λ under $<^*_{\sigma,\tau}$. Then j" is clearly α-recursive. The lemma is proved by showing that $j(n) = \lim_{\sigma\to\alpha} \lim_{\tau\to\alpha} j''(\sigma,\tau,n)$.

We make the following claim:

Claim For each $\nu < \lambda$, there is a $\gamma > h(\nu)$ and an $n(\nu)$ such that for all $\sigma > \gamma$, $\lim_{\tau\to\alpha} j''(\sigma,\tau,n(\nu) = \nu$.

Pick the least $\gamma \in \alpha - (M \cup B_\nu)$ such that $\gamma > h(\nu)$. As in the proof of the claim in Lemma 4.21, there are at most finitely many ordinals $\mu < \lambda$ such that $h'(\gamma+1,\mu) < \gamma+1$ (for otherwise $\gamma \in B_\mu$ for infinitely many μ, yielding an infinite descending sequence of ordinals). Let τ^* and μ_i, i = 1,...,n be chosen so that $\gamma \in B_{\mu_i}^{\tau^*}$ and $\gamma \notin B_\mu$ for $\mu \neq \mu_i$, for each i. Then $h''(\gamma+1,\tau^*, \mu_i) < h''(\gamma+1,\tau^*,\nu)$, i = 1,...,n. Also, for $\mu \neq \mu_i, \nu$, we have $j^{-1}(\mu)$ to be greater than $j^{-1}(\nu)$ (since $\gamma \notin B_\mu$). Hence the order type of ν under $<^*$ is to be decided among the ordinals $\nu, \mu_1,\ldots,\mu_n$. Now for each i, if $\mu_i <^* \nu$, then either $h(\mu_i) < h(\nu)$ or $h(\mu_i) = h(\nu)$

and $\mu_i < \nu$. Without loss of generality, let us suppose that $\mu_i <^* \nu$ for each $i = 1,\ldots,n$. Let $K = \{i \mid h(\mu_i) < h(\nu)\}$. Then there is a stage $\tau_1 > \tau^*$ such that $h(\mu_i) = h''(\gamma+1), \tau_1,\mu_i) < h''(\gamma+1,\tau_1,\nu)$ for all $i \in K$ (by admissibility and by the fact that K is a finite set). Then for all $\tau > \tau_1$, $j''(\gamma+1,\tau,j^{-1}(\nu)) = \nu$. Now consider $\sigma > \gamma$ in general. If there is no $\delta \in \alpha - M$ belonging to $\{x \mid \gamma < x < \sigma\}$, then clearly

$$j''(\sigma,\tau, j^{-1}(\nu)) = j''(\gamma+1,\tau,j^{-1}(\nu)) = \nu$$

for $\tau > \tau_1$, as above. If, however, such a δ exists, we pick the least one among them and apply the same argument as for γ earlier. Thus at most finitely many ordinals $\mu_1,\ldots,\mu_n$ may exist with $\delta \in B_{\mu_i}$, $i = 1,\ldots,n$. Let $j(i),\ldots,j(k)$ be chosen from $1,\ldots,n$ to be all those i's where $h(\mu_{j(i)}) < h(\nu)$. Then arguing as before there is a τ_2 such that $j''(\sigma,\tau,j^{-1}(\nu)) = \nu$ for all $\tau > \tau_2$. This proves the claim.

Lemma 4.23 now follows immediately from the claim.

Thus there is an $S_3(L_\alpha)$ function from ω onto λ, and by Lemma 4.19 there is an $S_2(L_\alpha)$ function from λ onto α. The composition of these two functions gives an $S_3(L_\alpha)$ function ω onto α. The proof of Theorem 4.17 is completed.

Finally, we take up the case where α has a maximal set with bounded complement. This is a situation unique to α-recursion theory. In this case one arrives at a conclusion similar to that of Theorem 4.17. However, it turns out that something much stronger is true. We first prove a lemma.

4.25 LEMMA Let W be an α-r.e. set bounded by $\gamma < \alpha$. If the order type of $\gamma - W$ is less than α^*, then $\gamma - W$ is α-finite.

Proof. By Lemma 4.20 there is a stage σ such that order type $(\gamma - W)$ = order type $(\gamma - W^\sigma)$. If $\alpha^* = \alpha$ then every bounded α-r.e. set is α-finite, and so we are done. If $\alpha^* < \alpha$ then α^* is the greatest α-cardinal and so, by the fact that $\gamma - W$ has order type less than α^*, it has α-cardinality less than α^* as well. Choose the stage σ as above. Then $(\gamma - W) \subseteq (\gamma - W^\sigma)$, and the latter is α-finite of α-cardinality less than α^*. Then $K = W \cap (\gamma - W^\sigma)$ is a bounded α-r.e. set of α-cardinality less than α^*. It is enough to show that K is α-finite. Let $t: \alpha \to K$ be a one-one α-recursive enumeration. The domain of t must be a proper initial segment of α, for otherwise $t''\alpha^*$ is a subset of K and has α-cardinality α^*, a contradiction. Let ν be the least ordinal for which $t(\nu)$ is not defined. Then $K = t''\nu$ and so it is α-finite. This implies that $(\gamma - W) = (\gamma - W^\sigma) - K$ is α-finite.

<u>4.26 THEOREM</u> If there is a maximal α-r.e. set M with bounded complement, then $\alpha > \alpha^* = \omega$.

Proof. It is immediate that $\alpha > \alpha^*$ since otherwise $\alpha - M$ will be finite. Suppose that $\alpha - M \subseteq \gamma$ for some $\gamma < \alpha$. Since α^* is the greatest α-cardinal (Lemma 2.8), there is an α-finite bijection p of γ with α^*. Let $M' = \alpha - p''[\alpha - M]$. We claim that M' is maximal. Clearly $\alpha - M'$ is not α^*-finite. Let W' be α-r.e. We may assume that $W' \subset \alpha^*$. Let $W = (p^{-1})''W'$. By the maximality of M, either $W - M$ is α^*-finite or $\alpha - (W \cup M)$ is α^*-finite. Let the resulting α^*-finite set be K. Then K is α-finite and of order type less than α^*. Clearly $p''K$ is α-finite. Furthermore it has α-cardinality less than α^*, since α^* is an α-cardinal. Thus it also has order type less than α^*. Now depending on the case that K falls under, either $p''K = W' - M'$ or $p''K = \alpha - (W' \cup M') = \alpha^* - (W' \cup M')$. This shows that M' is maximal.

We may thus assume that M is a maximal set whose complement is a subset of α^*.

Now notice that the order type of $\alpha^* - M$ is α^*. For otherwise by Lemma 4.25 $\alpha^* - M$ hence $\alpha - M$ is α-finite, contradicting Proposition 4.3.

We next split α^* into the union of a simultaneous α-r.e. sequence of α-r.e. sets A_ν, $\nu < \alpha^*$, such that the A_ν's are pairwise disjoint and each with order type α^*. This can be easily done. Define as in the unbounded case,

$$B_\nu = \{\gamma \mid (\exists\sigma)[\text{order type } \gamma - M^\sigma \in A_\nu]\}.$$

By our assumption that $\alpha - M \subset \alpha^*$, we see that $B_\nu \subset \alpha^*$ as well. Now each B_ν is α-r.e. and by Lemma 4.20 $B_\nu - M$ has order type α^*. By the maximality of M, $\alpha^* - (M \cup B_\nu)$ is α^*-finite, i.e. α-finite and of order type less than α^*. Once again by Lemma 4.20 there is a stage σ where the order of $\alpha^* - (M^\sigma \cup B_\nu^\sigma)$ is less than α^*. The map $h\colon \alpha^* \to \alpha$ such that $h(\nu)$ is the least stage σ where the above holds is α-recursive. Since $\alpha^* < \alpha$ we know that $h''\alpha^*$ is bounded in α so that h is in fact an α-finite function. Let $ot(\nu)$ be the order type of the set $\alpha^* - (M^{h(\nu)} \cup B_\nu^{h(\nu)})$. Again $ot\colon \alpha^* \to \alpha^*$ is α-finite. We use this to show that $\alpha^* = \omega$. Suppose $\alpha^* > \omega$.

<u>Claim</u> There is a $\delta < \alpha^*$ such for infinitely many $\nu < \alpha^*$ we have $ot(\nu) < \delta$.

If for each $\delta < \alpha^*$, there are only finitely many $\nu < \alpha^*$ such that $ot(\nu) < \delta$, then the α-cardinality of α^* is ω, which is not possible unless $\alpha^* = \omega$ (since α^* is itself an α-cardinal). The claim follows. Fix a δ as in the claim. Let

$$K = \{\nu \mid \nu < \alpha^* \ \& \ (\exists n)[\nu = \text{the } n\text{th ordinal } \sigma \text{ such that } ot(\sigma) < \delta\}.$$

Set

$$K' = \bigcup (\alpha^* - (M^{h(\nu)} \cup B_\nu^{h(\nu)})),\ \nu \in K.$$

Now since K is an α-finite set of order type ω and for each $\nu \in K$, the set $\alpha^* - (M^{h(\nu)} \cup B_\nu^{h(\nu)})$ is an α-finite set of order type less than δ, we see that K' is in fact an α-cardinality less than α^*. Since the order type of $\alpha^* - M$ is α^*, there exists an element γ in $\alpha^* - (M \cup K')$, i.e. γ is in B_ν for all $\nu \in K$. Now K is infinite, so this says that there exist infinitely many stages σ where $\alpha - M^\sigma$ assume different order types. As in the unbounded case this implies the existence of an infinite descending sequence, which is impossible. Thus $\alpha^* = \omega$ and this proves Theorem 4.26.

The converse of Theorem 4.26 is also true, namely there is a maximal α-r.e. set M with bounded complement if $\alpha > \alpha^* = \omega$. The construction essentially goes back to Kreisel and Sacks [1965]. We will not discuss it here.

CHAPTER 5

MAJOR SUBSETS

We continue with the study of the e-state method. This is carried out in the investigations of the structure $E(\alpha)$, the lattice of α-r.e. sets under inclusion. More precisely, we consider the problem of the existence of a major subset of a non-α-recursive α-r.e. set. This was first studied in Lachlan [1968] in connection with some problems related to the decidability of the theory of $E(\omega)$. Lerman [1976] introduced the notion of a major subset into α-recursion theory with a view to studying the 1-types of simple α-r.e. sets for various α. Since maximal sets do not exist for many α (Chapter 4), it was hoped that the structure of simple sets for such α might be easier to analyze. The study of major subsets was a step in that direction. As it turned out, however, the existence of major subsets is not a phenomenon shared by all admissible ordinals. Nevertheless, the results of Lerman and Leggett-Shore [1976] on major subsets do imply that for every admissible ordinal there exist at least two different 1-types for the class of simple α-r.e. sets. This information should prove useful in later studies.

In this chapter we give a necessary and sufficient condition on α for every non-α-recursive α-r.e. set to have a major subset. This time the condition hinges on the relative sizes of $\sigma 2cf(\alpha)$ and $\sigma 2p(\alpha)$. It shows once again the inherent limitations of many constructions in classical recursion theory. The first result on existence was proved in Lerman [1976] where it was shown that if α is a regular cardinal in L, then every non-α-recursive α-r.e. set has a major subset. Leggett and Shore [1976] extended this result to all α such that $\sigma 2cf(\alpha) >$

$\sigma 2p(\alpha)$. Chong [1979a] showed that this is the best possible, in the sense that counterexamples exist whenever $\sigma 2cf(\alpha) < \sigma 2p(\alpha)$. In particular, if $\sigma 2cf(\alpha) < \sigma 2p(\alpha)$, then no hyperhypersimple set has a major subset.

We begin with a lemma for arbitrary α-r.e. sets.

<u>5.1 LEMMA</u> Let A be an α-r.e. set which is not α-recursive. Let $\beta \leq \alpha$ be the least ordinal γ such that $A|\gamma$ is not α-recursive. Then $\sigma 2cf_\alpha(\beta) = \sigma 2cf(\alpha)$.

Proof. By the assumption on β, we know that $A|\gamma$ is α-recursive (hence α-finite) for all $\gamma < \beta$. Hence there exists a stage $s(\gamma)$ such that $A|\gamma = A^{s(\gamma)}|\gamma$ for every $\gamma < \beta$ (as usual A^σ is the set of elements enumerated in A by the end of stage σ). Let $f\colon \sigma 2cf(\alpha) \to \alpha$ be a $\Sigma_2(L_\alpha)$ cofinality function. Let R be a $\Sigma_2(L_\alpha)$ relation defined so that

$$R(i,\gamma) \quad \leftrightarrow \quad A^{f(i)}|\gamma \neq A|\gamma,$$

for $i < \sigma 2cf(\alpha)$ and $\gamma < \beta$. By the Σ_2 Uniformization Theorem (Theorem 1.27) there is a $\Sigma_2(L_\alpha)$ function u such that $R(i,u(i))$ holds for all $i < \sigma 2cf(\alpha)$. Furthermore u is cofinal in β since $A|\beta$ is not α-recursive and for $\gamma < \beta$, $A|\gamma$ is enumerated by stage $s(\gamma)$. It follows that $\sigma 2cf_\alpha(\beta) \leq \sigma 2cf(\alpha)$. Conversely, if u is a $\Sigma_2(L_\alpha)$ cofinality function from $\sigma 2cf_\alpha(\beta)$ into β, then $s \circ u\colon \sigma 2cf_\alpha(\beta) \to \alpha$ is a $\Sigma_2(L_\alpha)$ cofinal function. Hence $\sigma 2cf_\alpha(\beta) = \sigma 2cf(\alpha)$.

<u>5.2 REMARK</u> One may prove Lemma 5.1 directly without appealing to Theorem 1.27. We presented the proof as a (simple) illustration of how the theorem may be applied to admissible recursion theory.

Given two sets A and B, write $A \supseteq^* B$ if $B - A$ is α^*-finite.

As pointed out in Chapter 4, the class of α^*-finite sets is definable in the language of the lattice of α-r.e. sets.

5.3 DEFINITION Let A be an α-r.e. set. An α-r.e. set B contained in A is a major subset of A if A - B is not α^*-finite and if for all α-r.e. set C, whenever $C \supset^* (\alpha - A)$, then $C \supset^* (\alpha - B)$.

Thus if B is major subset of A, it is not possible to separate the complement of B from the complement of A using α-r.e. sets. Our construction of major subsets will be divided into two cases. The first one considers the situation where $\sigma 2cf(\alpha) = \sigma 2p(\alpha)$. This should also provide the motivation for the construction in the second case.

5.4 THEOREM Let $\sigma 2cf(\alpha) = \sigma 2p(\alpha)$. Then every α-r.e. set which is non-α-recursive has a major subset.

Proof. Let A be a non-α-recursive, α-r.e. set and let $\beta < \alpha$ be the least γ such that $A|\gamma$ is not α-recursive. Then $A|\gamma$ is α-finite for all $\gamma < \beta$ and by Lemma 5.1 we have $\sigma 2cf(\alpha) = \sigma 2cf_\alpha(\beta)$. Moreover, by Lemma 1.60 we have $t\sigma 2p(\alpha) = \sigma 2p(\alpha) = \sigma 2cf(\alpha)$. Thus we may choose $h: \sigma 2cf(\alpha) \to \alpha$ to be a tame $\Sigma_2(L_\alpha)$ surjection, approximated by an α-recursive function h'. Let

$$f'(\sigma,i) = \begin{cases} \min\ [(\alpha - A^\sigma) \cap (\alpha - W^\sigma_{h'(\sigma,i)}) \cap \beta], & \text{if defined} \\ \text{undefined} & \text{otherwise.} \end{cases}$$

Thus the function f' is α-recursive (one knows exactly when $f'(\sigma,i)$ is undefined). Furthermore, $\lim_{\sigma\to\alpha} f'(\sigma,i)$ is undefined if and only if $A \cup W_{h(i)} \supseteq \beta$ (observe that $\lim_{\sigma\to\alpha} f'(\sigma,i)$, if exists, identifies the least ordinal not in $A \cup W_{h(i)} \cap \beta$). Let A be given a one-one α-recursive enumeration such that the element enumerated in A at stage σ

is denoted a^σ. For $e < \sigma 2cf(\alpha)$, let $E(e,a^\sigma)$ be the e-state of a^σ defined as

$$E(e,a^\sigma) = \{i \mid i < e \ \& \ a^\sigma \in W^\sigma_{h'(\sigma,i)} \ \& \ a^\sigma < f'(\sigma,i)\}.$$

The e-states are ordered lexicographically.

We construct a major subset B of A by induction. To this end we define an α-recursive function $g'\colon \alpha \times \sigma 2cf(\alpha) \to \beta$ such that at stage σ, $g'(\sigma,i)$ may be considered to be the potential value of the ith element in the complement of B (thus $\alpha - B$ has order type $\sigma 2cf(\alpha)$). At the beginning of stage σ, set as usual $A^{<\sigma}$ to be the union of A^τ for $\tau < \sigma$. Set $g'(<\sigma,i) = \lim_{\tau \to \sigma} g'(\tau,i)$ if the limit exists, and let it be undefined otherwise.

Find the least $i < \sigma 2cf(\alpha)$, if exists, for which either $g'(<\sigma,i)$ is not defined or it is defined but $E(i,a^\sigma)$ has higher i-state than $E(i,g'(<\sigma,i))$ ('higher' under the lexicographical ordering). For this least i, let $g'(\sigma,i) = a^\sigma$. Set $g'(\sigma,j) = g'(<\sigma,j)$ for all j less than this i, and set $g'(\sigma,j)$ to be undefined otherwise. If, on the other hand, no such i exists, set $g'(\sigma,j) = g(<\sigma,j)$ for all j. Let

$$B^\sigma = A^\sigma - \{g'(\sigma,j) \mid g'(\sigma,j) \text{ is defined}\}.$$

This ends the construction at stage σ.

Let $B = \bigcup B^\sigma$, $\sigma < \alpha$. Clearly B is an α-r.e. subset of A. We show that it is a major subset of A.

<u>5.5 LEMMA</u> For each $i < \sigma 2cf(\alpha)$, $\lim_{\sigma \to \alpha} g'(\sigma,i) = g(i)$ exists.

Proof. This is proved by induction on i. Suppose that for all $i' < i$, $\lim_{\sigma \to \alpha} g'(\sigma,i') = g(i')$ exists. Observe that g is $\Sigma_2(L_\alpha)$ with domain $\sigma 2cf(\alpha)$. Since $i < \sigma 2cf(\alpha)$, the supremum of $\{g(i') \mid i' < i\}$ exists and there is a stage σ_0 such that for all $\sigma > \sigma_0$ we have

$g'(\sigma,i') = g'(\sigma_0,i') = g(i')$ for all $i' < i$. Since $t\sigma 2p(\alpha) = \sigma 2cf(\alpha)$ and $h: \sigma 2cf(\alpha) \to \alpha$ is a tame $\Sigma_2(L_\alpha)$ surjection, there is a stage σ_1 where $h'(\sigma,i') = h'(\sigma_1,i') = h(i')$ for all $\sigma > \sigma_1$ and $i' < i$. Without loss of generality let us assume that $\sigma_1 > \sigma_0$. Now for stages $\sigma > \sigma_1$, we know that the i'-states $E(i',a^\sigma)$, for $i' < i$, is not greater than $E(i',g'(\sigma_1,i'))$, since $g(i')$ has attained its final position. Thus if a^σ has a higher i-state than $g'(<\sigma,i)$ then attempt will be made at stage σ to make $g'(\sigma,i) = a^\sigma$. This process will continue at each stage σ after σ_1 whenever it appears that a higher i-state can be achieved. Let

$$E(i) = \{i' \mid i' < i \ \& \ W_{h(i')} \cup A \supseteq \beta\}.$$

$E(i)$ is a $\Pi_2(L_\alpha)$ subset of i, hence bounded below $\sigma 2cf(\alpha) = \sigma 2p(\alpha)$. As $\sigma 2p(\alpha)\mid(i+1) - E(i) = E'(i)$ is Σ_2 and bounded in $\sigma 2p(\alpha)$, it is α-finite, and so we have $E(i)$ to be α-finite as well. For each i' in $E'(i)$, $f(i') = \lim_{\sigma\to\alpha} f'(\sigma,i')$ exists and is less than β. Once again since $i < \sigma 2cf_\alpha(\beta)$, there is a uniform upper bound $\zeta < \beta$ for the set

$$\{f(i') \mid i' < i \ \& \ i' \in E'(i)\}.$$

By our choice of the ordinal β, we know that $A\mid\gamma$ is α-finite for every $\gamma < \beta$. Thus there is an ordinal $\sigma_2 > \sigma_1$ such that $A^{\sigma 2}\mid\zeta = A\mid\zeta$. We proceed from stage $\sigma > \sigma_2$. Consider $g'(\sigma_2,i)$. If this ordinal exists and has i-state not less than the i-state of any future a^σ, $\sigma > \sigma_2$, then clearly $\lim_{\sigma\to\alpha} g'(\sigma,i) = g'(\sigma_2,i)$ and we are done. Thus assume that there are stages $\tau > \sigma$ where the i-state of a^τ is greater than the i-state of $g'(\sigma,i)$. The construction then stipulates that $g'(\tau,i)$ be set to be equal to a^τ. We claim that this process cannot go on unboundedly often in α. Let $D(i)$ be the α-r.e. set $\bigcap W_{h(i')} \cap \beta$, for i' in $E(i)$. Then $D(i) \cup A \supset \beta$. If there is an a^σ in $D(i) \cap A$ such that $a^\sigma < f'(\sigma,i')$ for all i' in $E(i)$, then clearly it has the greatest i-state at stage σ and so we would set $g'(\sigma,i) = a^\sigma$, ensuring by the maximality of the i-state of a^σ that $g'(\tau,i) = g'(\sigma,i)$ for all $\tau > \sigma$. This would prove that $g(i) = \lim_{\sigma\to\alpha} g'(\sigma,i)$ exists. We show that such an a^σ does indeed exist (in fact as will be seen from the argument below the set of such a^σ's is not α-finite). For suppose that this is not the case. The following gives an algorithm for computing A in an α-recursive way. To obtain $A|\gamma$, for $\gamma > \zeta$, go to a stage σ greater than σ_2 where $A^\sigma \cup (D(i))^\sigma \supset \gamma$. Then no element less than γ which has not entered A by stage σ will ever enter A. This is true since if a^τ is so enumerated at stage τ, then one has $a^\tau \in A \cap D(i)$ and a^τ is less than $f'(\tau,i')$ for every $i' \in E(i)$, contradicting our assumption. Hence we get $A^\sigma|\gamma = A|\gamma$, and so A is α-recursive. This contradiction proves Lemma 5.5.

<u>5.6 LEMMA</u> The set $X = \{g(i)\,|\,i < \sigma 2cf(\alpha)\}$ is not contained in any A^σ. Hence it is not α-finite.

Proof. Suppose that $X \subseteq A^\sigma$. Then there is a stage σ' such that $g(i) = g'(\tau,i)$ for all $\tau > \sigma'$ and all $i < \sigma 2cf(\alpha)$. This means that the function g is α-recursive. Now since g' is defined in terms of h', we have also $h'(\tau,i) = h(i)$ for all $\tau > \sigma'$ and all $i < \sigma 2cf(\alpha)$. This means that the function h is α-recursive. Since h is a surjection from $t\sigma 2p(\alpha) = \sigma 2cf(\alpha)$ onto α, we must have $\sigma 2cf(\alpha) = \alpha$. But now g is an α-recursive injection from α into $\sup A^\sigma < \alpha$. This implies that $\alpha^* < \alpha$ and so $\sigma 2cf(\alpha) < \alpha$. This gives a contradiction and so we conclude that X is not contained in any A^σ.

Observe that $B = A - X$, where X is as defined in Lemma 5.6. We now show that B is a major subset of A.

<u>5.7 LEMMA</u> Suppose that C is α-r.e. and $C \supset^* (\alpha - A)$. Then $C \supset^* (\alpha - B)$.

Proof. By the definition of B we may assume without loss of generality that the set C covers the complement of B in the set $\alpha - \beta$ (since C covers, with the possible exception of α^*-finitely many elements, the complement of A in $\alpha - \beta$). Let $C = W_{h(i)}$. We show that $g(j) \in C$ for all j sufficiently large. Let E(i) and E'(i) be as defined in the proof of Lemma 5.5 and let D(i) be the corresponding α-r.e. set. Also let σ^* be an ordinal chosen satisfying the following properties: $g'(\sigma^*,i') = g(i')$ for all $i' < i$; $h'(\sigma,i') = h(i')$ for all $\sigma > \sigma^*$ and $i' < i$; and $\lim_{\sigma \to \alpha} f'(\sigma,i') = f(i') < a^{\sigma}$ for all i' in E'(i) and $\sigma > \sigma^*$. Now by Lemma 5.6, the set X defined there is not contained in A^{σ^*}. Let j_0 be the least j greater than i such that $g(j) \not\in A^{\sigma^*}$. Fix $j > j_0$. By the choice of σ^*, we have $g'(\sigma,j) > f'(\sigma,i')$ for each $i' \in E'(i')$ and $\sigma > \sigma^*$. Now as argued in Lemma 5.5 there exist α-infinitely many stages $\sigma > \sigma^*$ such that $a^{\sigma} \in D(i)$ and $a^{\sigma} < f'(\sigma,i')$ for each i' in E(i). For each such a^{σ}, if there is a $\tau > \sigma$ such that $E(j,a^{\tau}) > E(j,a^{\sigma})$, then one must also have $a^{\tau} \in D(i)$ and $a^{\tau} < f'(\tau,i')$ for all $i' \in E(i)$ (by the lexicographical ordering of the j-states). This means that in our attempt to make g(j) attain a higher j-state, we would actually have to choose it from among those elements in D(i). In particular, $g(j) \in W_{h(i)}$. Since this holds for all $j > j_0$ we have $W_{h(i)} = C \supset^* (\alpha - B)$.

This concludes the proof of Theorem 5.4.

We next discuss the case when $\sigma 2p(\alpha) < \sigma 2cf(\alpha)$. Now by Theorem 1.28 there is a $\Sigma_2(L_\alpha)$ function (again denoted h) mapping a subset of $\sigma 2p(\alpha)$ onto α. By the fact that $\sigma 2p(\alpha) < \sigma 2cf(\alpha)$, this subset is not α-r.e., since otherwise $\sigma 2cf(\alpha) \leq \alpha^* \leq \sigma 2p(\alpha)$, a contradiction. Thus the natural priority given above for the case $\sigma 2p(\alpha) = \sigma 2cf(\alpha)$, natural in the sense that the ith requirement (saying if $W_{h(i)} \supseteq^* (\alpha - A)$ then $W_{h(i)} \supseteq^* (\alpha - B)$) has higher priority than the jth requirement if $h(i) < h(j)$, is no longer useful here since i may not be in the domain of h, so that $\lim_{\sigma \to \alpha} h'(\sigma,i)$ may not exist (h' is the α-recursive approximation of h), even though convergence may appear to hold for arbitrarily long intervals of stages for $h'(\sigma,i)$.

The way out of this difficulty is to introduce a new ordering on the requirements based on the length of the period of convergence. More precisely, at stage σ let

$$k'(\sigma,i) = \text{the least } \tau \leq \sigma \text{ such that for all } \rho,\ \tau \leq \rho \leq \sigma, \text{ we have } h'(\rho,i) = h'(\tau,i)$$

Say that the ith requirement has higher priority than the jth requirement at stage σ, written $i <_\sigma j$, if $k'(\sigma,i) < k'(\sigma,j)$, or if $k'(\sigma,i) = k'(\sigma,j)$ and $i < j$. This gives an α-recursive well-ordering of $\sigma 2p(\alpha)$ of length less than $(\sigma 2p(\alpha))^+$, the next α-cardinal after $\sigma 2p(\alpha)$ (if this exists; otherwise set it to be α). Let $t(\sigma)$ be the order type of the ordering at stage σ. Observe that if $i <_\sigma j$, then at stage σ we know that either the convergence to $h'(\sigma,i)$ occurred at a stage before the convergence to $h'(\sigma,j)$, or the convergence took place at the same stage but $i < j$. This gives a way of comparing how 'long' the 'correct values' (at stage σ) of h(i) and h(j) have been arrived at. We set $i = b(\sigma,\nu)$ if i the the νth element under this ordering (note that $\nu < t(\sigma)$).

Consider now $k(i) = \lim_{\sigma \to \alpha} k'(\sigma,i)$ for each i in the domain of h. Clearly k(i) is defined and is in fact equal to the least σ where $h'(\sigma,i) = h'(\tau,i)$ for all $\tau \geq \sigma$. For $\sigma' > \sigma \geq k(i)$, if $i = b(\sigma,\nu) = b(\sigma',\nu')$, then one has $\nu' \leq \nu$. In other words, after stage k(i), the order of i under $<_\sigma$ can at best decrease as σ increases. Since every decreasing sequence of ordinals is finite, there is a stage $\sigma(i) \geq k(i)$ where for all $\sigma' > \sigma \geq \sigma(i)$, $i = b(\sigma',\nu) = b(\sigma,\nu)$ for some fixed ν. On the other hand, given i and j in the domain of h, we may define $i <_h j$

if either $k(i) < k(j)$, or if $k(i) = k(j)$ and $i < j$. This gives a well-ordering on the domain of h. Let t be the order type of $<_h$. Let $i = b_\nu$ if i the the νth element under $<_h$. What we now have is a correspondence: for each ν, $b_\nu = \lim_{\sigma\to\alpha} b(\sigma,\nu)$; moreover,

(5.8) the ordinal $t < (\sigma 2p(\alpha))^+$, and

(5.9) for each $\nu < t$, there is a stage σ such that $b_\mu = b(\tau,\mu)$ for all $\tau > \sigma$ and $\mu < \nu$.

We first prove (5.8). Let $i \in \mathrm{dom}(h)$. Let

$$K = \{j \mid j < \sigma 2p(\alpha) \ \& \ (\exists\sigma > k(i))[h'(\sigma,j) \neq h'(k(i),j)]\}.$$

Then K is an α-r.e. subset of $\sigma 2p(\alpha)$. K is α-finite since otherwise $\alpha^* < \sigma 2p(\alpha)$, which would imply that $\sigma 2cf(\alpha) < \sigma 2p(\alpha)$, contradicting our assumption. It follows that $\sigma 2p(\alpha) - K$ is also α-finite, and that every j in $\sigma 2p(\alpha) - K$ is in the domain of h, with the property that $k(j) < k(i)$. Thus $\{j \mid j <_h i\} = \{j \mid j \in \sigma 2p(\alpha) - K\}$. Now for such j, $j <_h i$ if and only if $j <_{k(i)} i$. The latter is an α-recursive relation. It follows that the map $j \to \mu$ where $j = b_\mu$ is α-finite taking $\sigma 2p(\alpha) - K$ into an initial segment of α. Now $(\sigma 2p(\alpha))^+$ is either an α-cardinal or equal to α. Since no α-finite map may take an α-finite set bounded by $\sigma 2p(\alpha)$ onto $(\sigma 2p(\alpha))^+$, we know that if $i = b_\nu$, then $\nu < (\sigma 2p(\alpha))^+$. Since i is arbitrary, this proves (5.8).

Now (5.9) follows from the proof of (5.8). Suppose that $i = b_\nu$. Then as noted above the set K is α-finite and for $j \in \sigma 2p(\alpha) - K$, $j <_h i$ if and only if $j <_{k(i)} i$. For such a j, we have $j = b(k(i),\mu)$ for some $\mu < \nu$. Furthermore, let $\sigma > k(i)$ be chosen such that every element of K has been enumerated by stage σ. Then for all $\tau > \sigma$, $\mu < \nu$, if $j = b(\tau,\mu)$ then $j = b_\mu$.

We next consider pairs (δ,K) where $\delta < \sigma 2p(\alpha)$ and K is a non-empty α-finite subset of δ. Fix a stage σ. We say that (δ,K) is a candidate at stage σ if there is a $\nu < t(\sigma)$ such that

$$K = \{i \mid i < \delta \ \& \ (\exists\mu<\nu)(i = b(\sigma,\mu)\}.$$

In other words (δ,K) is a candidate at stage σ if K is the restriction to δ of a proper initial segment of ordinals ordered under $<_\sigma$. From now on we consider only pairs (δ,K) which are candidates. Let $\nu(\delta,K,\sigma)$ be the least ν in the above definition. We introduce the following priority system: (δ,K) has higher priority than (ζ,N) at stage σ if either $\nu(\delta,K,\sigma) < \nu(\zeta,N,\sigma)$, or if these two ordinals are equal and $\delta < \zeta$. The order type of this priority system (at stage σ) is $\sigma 2p(\alpha).t(\sigma)$, which is again less than $(\sigma 2p(\alpha))^+$. If (δ,K) is the νth element under this ordering at stage σ, we write $P'(\sigma,(\delta,K)) = \nu$, and say that the priority of (δ,K) at stage σ is ν. If $\lim_{\sigma\to\alpha} P'(\sigma,(\delta,K)) = P(\delta,K)$ exists, we say that the priority of (δ,K) is $P(\delta,K)$. Now observe the following:

(5.10) If $K \subset \mathrm{dom}(h) \cap \delta$, then it follows from (5.9) that $\lim_{\sigma\to\alpha} \nu(\delta,K,\sigma)$ exists, which in turn implies that $\lim_{\sigma\to\alpha} P'(\sigma,(\delta,K)) = P(\delta,K))$ exists. Conversely, if $P(\delta,K)$ exists, then $\lim_{\sigma\to\alpha} \nu(\delta,K,\sigma)$ exists and so $K \subset \mathrm{dom}(h) \cap \delta$.

(5.11) The priority system has order type $\sigma 2p(\alpha).t <$ $(\sigma 2p(\alpha))^+$ in the limit (i.e. as $\sigma \to \alpha$). This is a consequence of the fact that $t = \lim_{\sigma\to\alpha} t(\sigma)$.

(5.12) Again it follows from (5.9) that for all ν, there is a stage σ_0 such that $P'(\sigma,(\delta,K)) = P(\delta,K)$ for all $\sigma > \sigma_0$ and all (δ,K) such that (δ,K) has higher priority than ν (in the limit).

We are now ready for the next result.

<u>5.13 THEOREM</u> Suppose that $\sigma 2p(\alpha) < \sigma 2cf(\alpha)$. Then every non-$\alpha$-recursive α-r.e. set has a major subset.

Proof. Let A be a non-α-recursive α-r.e. set and as in the proof of Theorem 5.4 let $\beta \leq \alpha$ be the least ordinal γ such that $A|\gamma$ is not α-recursive. By Lemma 5.1 we have $\sigma 2cf(\alpha) = \sigma 2cf_\alpha(\beta)$. Give A a one-one α-recursive enumeration so that at stage σ the element a^σ is enumerated. Define the α-recursive function $f'\colon \alpha\times\sigma 2p(\alpha) \to \beta$ as before. If $i \in dom(h)$, then either $\lim_{\sigma\to\alpha} f'(\sigma,i) < \beta$, in which case we know $A \cup W_{h(i)}$ does not contain β, or $\lim_{\sigma\to\alpha} f'(\sigma,i) = \beta$, in which case we have $A \cup W_{h(i)} \supset \beta$. The point to be stressed here is that if i is not in the domain of h, then $\lim_{\sigma\to\alpha} h'(\sigma,i)$ does not exist, and so neither does $\lim_{\sigma\to\alpha} f'(\sigma,i)$. On the other hand, if $i \in dom(h)$, then for stages $\sigma > k(i)$, $f'(\sigma,i)$ is non-decreasing, and reaches a limit below β if and only if $A \cup W_{h(i)} \not\supset \beta$. Define the K-state of a^σ to be the set

$$E(K,a^\sigma) = \{i | i \in K \;\&\; a^\sigma \in W_{h'(\sigma,i)} \;\&\; a^\sigma < f'(\sigma,i)\}.$$

Order the K-states lexicographically. Observe that if K is the set of ordinals less than e, then $E(K,a^\sigma) = E(e,a^\sigma)$ as defined in the proof of Theorem 5.4. We define an α-recursive function g' on pairs (δ,K). We ensure that $\lim_{\sigma\to\alpha} g'(\sigma,(\delta,K)) = g(\delta,K)$ exists if and only if K is contained in the domain of h. The set $\{g(\delta,K)|g(\delta,K)$ is defined$\}$ is a subset of A of order type $\sigma 2p(\alpha).t$, and is the set A - B.

At stage σ, set $g'(<\sigma,(\delta,K)) = \lim_{\tau\to\sigma} g'(\tau,(\delta,K))$ if the limit exists and $\lim_{\tau\to\sigma} P'(\tau,(\delta,K)) = P'(\sigma,(\delta,K))$. Let it be undefined otherwise. Look for the pair (δ,K) which has the highest priority at stage σ (thus $P'(\sigma,(\delta,K))$ is defined) such that either $g'(<\sigma,(\delta,K))$ is not defined, or is defined but $E(K,a^\sigma)$ has higher K-state than $E(K,g'(<\sigma,(\delta,K))$. Denote this by (δ_0,K_0). Define $g'(\sigma,(\delta,K)) = g'(<\sigma,(\delta,K))$ if (δ,K) has higher priority than $(\delta 0,K0)$ at stage σ. Set

$g'(\sigma,(\delta_0,K_0)) = a^\sigma$ if (δ_0,K_0) exists. Let $g'(\sigma,(\delta,K))$ be undefined for the other (δ,K). If (δ_0,K_0) does not exist, this means that no change in K-state is necessary, and so we set $g'(\sigma,(\delta,K)) = g'(<\sigma,(\delta,K))$ for every (δ,K). Let

$$B^\sigma = A^\sigma - \{g'(\sigma,(\delta,K)) \mid g'(\sigma,(\delta,K)) \text{ is defined}\}.$$

Set $B = \bigcup B^\sigma$, $\sigma < \alpha$. This completes the construction of the set B. We show that B is a major subset of A.

5.14 LEMMA For each pair (δ,K), if $P(\delta,K)$ is defined (so that by (5.10) $K \subset \mathrm{dom}(h)$), then $\lim_{\sigma\to\alpha} g'(\sigma,(\delta,K)) = g(\delta,K)$ exists.

Proof. We prove this by induction on the priority of (δ,K). Suppose that this is true of all pairs of higher priority than (δ,K) (in the limit). By (5.12) choose a stage σ_0 such that $P'(\sigma,(\zeta,N)) = P(\zeta,N)$ for all $\sigma > \sigma_0$ and all (ζ,N) with $P(\zeta,N) < P(\delta,K)$. The map $(\zeta,N) \to g(\zeta,N)$, for $P(\zeta,N) < P(\delta,K)$, is then a $\Sigma_2(L_\alpha)$ function defined on the initial segment of ordinals less than $P(\delta,K)$ with image in β. As $P(\delta,K) < (\sigma 2p(\alpha))^+ < \sigma 2cf(\alpha)$, and $\sigma 2cf_\alpha(\alpha) = \sigma 2cf(\alpha)$ (Lemma 5.1), the image of g defined on this initial segment is bounded below β. In particular there is a stage $\sigma_1 > \sigma_0$ such that for all $\sigma > \sigma_1$ we have $g'(\sigma,(\zeta,N)) = g'(\sigma_1,(\zeta,N))$. Consider now the stages after σ_1. For $\sigma > \tau > \sigma_1$, $g'(\sigma,(\delta,K)) \neq g'(\tau,(\delta,K))$ if and only if the K-state of the former is greater than that of the latter. Let

$$E(\delta,K) = \{i \mid i \in K \ \& \ A \cup W_{h(i)} \supseteq \beta\}.$$

Observe that $h(i)$ is defined for each i in K. Consider now the set $E'(\delta,K) = K - E(\delta,K)$. By the choice of the stage σ_1, we have $P(\zeta,N)$ equal to $P'(\sigma_1,(\zeta,N))$ for every (ζ,N) of higher priority than (δ,K). Since every proper initial segment N of K may be associated with a ζ to form a pair (ζ,N) of higher priority than (δ,K), this in turn implies that $h'(\sigma,i) = h(i)$ for all $i \in K$ and all $\sigma > \sigma_1$ (since the priority function P' is defined in terms of k' which is itself defined in terms of h'). It follows that $E'(\delta,K)$ is a $\Sigma_2(L_\alpha)$ subset of K bounded below $\sigma 2p(\alpha)$, hence α-finite. This implies that $E(\delta,K)$ is α-finite as well. The rest of the argument now proceeds as in Lemma 5.5, where the i-states are now replaced by K-states. We leave the verification to the reader.

Notice that if $g(\delta,K)$ is defined, then $P(\delta,K)$ must be defined (according to the construction of the set B), so that by (5.10) the set K is contained in the domain of h. Furthermore,

$$A - B = \{g(\delta,K) \mid g(\delta,K) \text{ is defined}\}$$
$$= \{g(\delta,K)) \mid \delta < \sigma 2p(\alpha) \ \& \ K \subset \mathrm{dom}(h)\}.$$

This equality follows from the construction at stage σ, since if it is found at that stage that a^σ has a higher K-state than $g'(<\sigma,(\delta,K))$ for some (δ,K), then a^σ is put into B. The elements of A that stay out of B permanently are the ones which are equal to $g'(\sigma,(\delta,K))$, for some (δ,K), at all sufficiently large σ.

<u>5.15 LEMMA</u> The set $X = \{g(\delta,K) \mid \delta < \sigma 2p(\alpha) \ \& \ K \subset \mathrm{dom}(h)\}$ is not contained in any A^σ, so that $A - B$ is not α^*-finite.

Proof. Suppose that $X \subset A^\sigma$. This means that $g(\delta,K) = g'(\sigma,(\delta,K))$ for all $\delta < \sigma 2p(\alpha)$ and $K \subset \mathrm{dom}(h)$. So at stage σ, $P'(\sigma,(\delta,K)) = P(\delta,K)$ for each such (δ,K). By the way the priority function P' was defined, it means that $h'(\sigma,i) = h(i)$ for all i in the domain of h. This gives rise to an α-recursive surjection $i \to h'(\sigma,i)$ from a subset of $\sigma 2p(\alpha)$ onto α. Hence $\sigma 2p(\alpha) = \alpha = \sigma 2cf(\alpha)$, a contradiction.

<u>5.16 LEMMA</u> If C is α-r.e. and $C \supset^* (\alpha - A)$, then we have

$C \supset^* (\alpha - B)$.

Proof. This is proved as in Lemma 5.7. Let $C = W_{h(i)}$. We may assume that C covers the complement of B in $\alpha - \beta$. Let

$$K = \text{dom}(h) \cap \{j \mid j < i\},$$

and let δ be $(\sup K)+1$. Define $E(\delta,K)$ and $E'(\delta,K)$ as in Lemma 5.14. Let $D(\delta,K) = \cap W_{h(i')}$, $i' \in E(\delta,K)$. Choose an ordinal σ_2 such that for all $\sigma > \sigma_2$ and (ζ,N) of higher priority than (δ,K), $g'(\sigma,(\zeta,N)) = g(\zeta,N)$ (hence also $P'(\sigma,(\zeta,N)) = P(\zeta,N)$), and $f(i') = \lim_{\sigma\to\alpha} f'(\sigma,i') = f'(\sigma_2,i')$ for each i' in $E'(\delta,K)$. By Lemma 5.15 the set X is not contained in any A^σ. Let (δ_0,K_0) be the pair of highest priority after (δ,K) whose image under g does not lie in $A^{\sigma 2}$. Then one argues, as in Lemma 5.7, that $g(\zeta,N)$ lies in $W_{h(i)} = C$ for each (ζ,N) whose priority is either equal to or lower than (δ_0,K_0). Thus at most the set

$$\{g(\zeta,N) \mid (\zeta,N) \text{ of higher priority than } (\delta_0,K_0)\}$$

is omitted in C. But this set is α-finite, since by the proof of Lemma 5.14 there is a stage τ where $g'(\sigma,(\zeta,N)) = g(\zeta,N)$ for all such (ζ,N) and all $\sigma > \tau$. This set is moreover of order type less than $(\sigma 2p(\alpha))^+ < \sigma 2cf(\alpha) < \alpha^*$. Thus $C \supset^* (\alpha - B)$.

This completes the proof of Theorem 5.13.

A consequence of Theorem 5.13 is the following fine structure result:

<u>5.17 COROLLARY</u> If $\sigma 2p(\alpha) < \sigma 2cf(\alpha)$, then $\sigma 2cf(\alpha) = (\sigma 2p(\alpha))^+$.

Proof. Let A be a non-α-recursive, α-r.e. regular set. Then the β defined in the proof of Theorem 5.13 is equal to α. Let B be a major subset of A as given in the construction. By Lemma 5.15 and the regularity of A the set A - B is not bounded in α. Thus A - B is $\Sigma_2(L_\alpha)$ and not α-finite and so must have order type at least $\sigma2cf(\alpha)$. For each $\gamma < \alpha$, $(A - B) \cap \gamma$ is α-finite (contained in some A^σ by the regularity of A) and is contained in a set of the form $\{g(\zeta,N) | P(\zeta,N) < P(\delta,K)\}$. The latter is an α-finite set of α-cardinality at most $\sigma2p(\alpha)$. Thus the set $(A - B) \cap \gamma$ has order type less than $(\sigma2p(\alpha)^+$. We now have $\sigma2cf(\alpha) \leq$ the order type of $(A - B) \leq (\sigma2p(\alpha))^+ \leq \sigma2cf(\alpha)$. Hence $\sigma2cf(\alpha) = (\sigma2p(\alpha))^+$.

Our next task is to show that the results of Theorem 5.4 and 5.13 are the best possible, in the sense that if $\sigma2cf(\alpha) < \sigma2p(\alpha)$, then not every non-$\alpha$-recursive α-r.e. set has a major subset. We will in fact show that if the hypothesis holds, then no α-r.e. set A with complement of order type less than $t\sigma2p(\alpha)$ (e.g. the set of non-constructible cardinals in $\aleph_\omega^{(L)}$) has a major subset. Thus we have here an example of an e-state construction which, as opposed to the one for maximal sets, fails not due to cardinality considerations, but to the lack of sufficient admissibility to carry through the priority argument. For Σ_2 inadmissible constructible cardinals α, we see that maximal sets do not exist and not every non-α-recursive α-r.e. set has a major subset. As we will see in Chapter 6, there is also a complete characterization of the hyperhypersimple r.e. sets on these α in terms of the order types of their complements. In Chapter 10 it will further be shown that the generalized version of Post's problem fails for many such α. All of these indicate strongly some unique features inherent in recursion theory on Σ_2 inadmissible cardinals of L. These matters merit further study.

For the rest of the chapter, let λ denote $\sigma2cf(\alpha)$.

<u>5.18 LEMMA</u> There is a uniformly α-r.e. sequence of pairwise disjoint α-r.e. sets $\{H_\nu | \eta < \lambda\}$ such that

(a) $\alpha = \bigcup\{H_\eta | \eta < \lambda\}$

(b) For all $\nu < \lambda$, $\bigcup\{H_\eta | \eta < \nu\}$ is α-finite.

Proof. Let f be a $\Sigma_2(L_\alpha)$ cofinality function from λ into α. Let f' be an α-recursive approximation of f in the usual sense. At stage σ, find the least $\zeta > \sigma$, denoted $\zeta(\sigma)$, such that $f'(\zeta,\eta)$ is not equal to $\lim_{\tau\to\zeta} f'(\tau,\eta)$ for some $\eta < \lambda$. Such a $\zeta(\sigma)$ exists sinced f' cannot approximate to the correct value for every $\eta < \lambda$ at any stage σ. Let $\eta(\sigma)$ be the least η such that

$$\lim_{\tau\to\zeta(\sigma)} f'(\tau,\eta) \neq f'(\zeta(\sigma),\eta).$$

Put σ into $H_{\eta(\sigma)}$. Then $\{H_\eta | \eta < \lambda\}$ is a uniformly α-r.e. sequence of pairwise disjoint α-r.e. sets. Also each σ enters some H_η at stage σ, so that (a) is true. Since f is Σ_2 cofinal, for each $\eta < \lambda$ there is a τ such that $f'(\sigma,\nu) = f'(<\sigma,\nu) = f(\nu)$ for all $\sigma > \tau$ and $\nu < \eta$. Then no ordinal σ greater than τ will enter any of H_ν, $\nu < \eta$. This implies (b).

<u>5.19 LEMMA</u> If A is an α-r.e. set whose complement in α has order type less than $t\sigma 2p(\alpha)$, then A is regular.

Proof. Observe that $t\sigma 2p(\alpha) \leq \alpha^*$. This follows from the fact that the inverse of the (α-recursive) projection $p_1: \alpha \to \alpha^*$ may be used to define a tame $\Sigma_2(L_\alpha)$ map g from α^* onto α. To see this, let $g'(\sigma,i)$ be 0 if $i \notin (p_1)''\sigma+1$, and otherwise let $g'(\sigma,i)$ be the ζ such that $p_1(\zeta) = i$ (recall that p_1 is one-one). Then g' is α-recursive and $\lim_{\sigma\to\alpha} g'(\sigma,i) = g(i)$ exists for each $i < \alpha^*$. Also g is tame since for each $\gamma < \alpha^*$, $\text{range}(p_1) \cap \gamma$ is α-finite. Thus the lemma would be proved once we have shown that A is regular whenever $\alpha - A$ has order type less than α^*. But this is a consequence of Lemma 4.25, since for each $\gamma < \alpha$, $A|\gamma$ is bounded in α and $(\alpha - A)|\gamma$ has order type less than α^*.

We leave the proof of the next lemma to the reader.

5.20 LEMMA There is a $\Sigma_2(L_\alpha)$ cofinality function $f\colon \lambda \to \alpha$ with an α-recursive approximation f' such that f is strictly increasing and $f'(\tau,\gamma) \leq f'(\sigma,\delta)$ for all $\tau \leq \sigma$ and $\gamma \leq \delta$.

5.21 THEOREM Let $\lambda < \sigma 2p(\alpha)$. Then no α-r.e. set A with $\alpha - A$ of order type less than $t\sigma 2p(\alpha)$ has a major subset.

The theorem is established through several lemmas. Let A be an α-r.e. set such that $\alpha - A$ has order type less than $t\sigma 2p(\alpha)$. Let B be an α-r.e. subset of A such that $A - B$ is not α^*-finite. We construct an α-r.e. set C such that $C \supset^* (\alpha - A)$ but not $C \supset^* (\alpha - B)$. Let γ_0 be the least γ for which $(A - B)|\gamma$ is not α^*-finite.

If $\gamma_0 < \alpha$, then by Lemma 5.20 we have $A|\gamma_0$ hence $(\alpha - A)|\gamma_0$ to be α-finite. Let C be an α-r.e. set which is the union of $\alpha - \gamma_0$ with $(\alpha - A)|\gamma_0$. Then $C \supset (\alpha - A)$ but $C \cap A = \emptyset$, so that certainly C does not contain element of $\alpha - B$. Thus B is not a major subset of A.

We may therefore assume from now on that the ordinal γ_0 is equal to α.

Let $\{H_\eta | \eta < \lambda\}$ and $f\colon \lambda \to \alpha$ be as given in Lemmas 5.18 and 5.20. Set $[\gamma,\delta) = \{\zeta | \gamma \leq \zeta < \delta\}$. Let

$$K^* = \{(\nu,\eta) | H_\eta \cap (\alpha - A) \neq \emptyset \ \& \ H_\eta \cap (A - B) \cap [f(\nu),\alpha) \neq \emptyset\}.$$

By Proposition 1.26, λ is an α-cardinal hence closed under the pairing function. In particular one may regard K^* to be a subset of λ. Then K^* is $\Sigma_2(L_\alpha)$ and bounded below $\sigma 2p(\alpha)$, hence α-finite. For each $\nu < \lambda$, let $k(\nu)$ be the least η such that $(\nu,\eta) \in K^*$. Let ν_0 be the least ν for which $k(\nu)$ is not defined. This means that for each $\eta < \lambda$, if there is a $\nu > \nu_0$ such that

$$H_\eta \cap (A - B) \cap [f(\nu),\alpha) \neq \emptyset,$$

then $H_\eta \cap (\alpha - A) = \emptyset$. Let

$$K' = \{\eta \mid (\exists\nu{>}\nu_0)[H_\eta \cap (A - B) \cap [f(\nu),\alpha) \neq \emptyset\}.$$

K' is a $\Sigma_2(L_\alpha)$ subset of λ and hence α-finite. Let $C = \bigcup\{H_\eta \mid \eta \notin K'\}$. Then $C \supset (\alpha - A)$ and $C \not\supset^* (\alpha - B)$. So B is not a major subset of A.

Thus let us assume that $k(\nu)$ is defined for every $\nu < \lambda$. Let $K = \{k(\nu) \mid \nu < \lambda\}$. For $\eta = k(\nu)$, let $r(\eta)$ be the least element of H_η (A - B) $\cap [f(\nu),\alpha)$. Then $R = \{r(\eta) \mid \eta < \lambda\}$ is a tame $\Sigma_2(L_\alpha)$ set of order type λ. For $\eta \in K$, let $\kappa(\eta)$ be the order type of the set

$$H_\eta \cap ((\alpha - A) \cap \{r(\eta)\}).$$

Let $r(\eta)$ be the $\rho(\eta)$th element (in magnitude) of this set. By the fact that α - A has order type less than $t\sigma 2p(\alpha)$, we see that there is an ordinal $s < t\sigma 2p(\alpha)$ such that $s > r(\eta)$ and hence $\rho(\eta)$ for each $\eta \in K$. We first prove a lemma.

<u>5.22 LEMMA</u> The maps $\eta \to \kappa(\eta)$ and $\eta \to \rho(\eta)$ are α-finite.

Proof. Define

$$S(\eta,z) \quad \leftrightarrow$$

$$(\exists t)[t\colon z \to H_\eta \ \&\ t \text{ is strictly increasing } \&$$
$$(\exists\sigma)(\forall\tau{>}\sigma)(\forall x{<}z)(t(x) \in (\alpha - A^\tau) \text{ or } t(x) = r(\eta)].$$

Since A is regular (Lemma 5.19) and R is tame $\Sigma_2(L_\alpha)$, we see that for each η there is a z such that $S(\eta,z)$ holds. Moreover there is a uniform upper bound less than $t\sigma 2p(\alpha)$ (indeed an ordinal not greater than s) for the set of such z's as η ranges over elements of K. Then S is a tame Σ_2 set and bounded below $\text{t}\sigma 2p(\alpha)$, hence α-finite. Observe that $\kappa(\eta)$ is the least z such that $\sim S(\eta,z+1)$. This shows that $\eta \to \kappa(\eta)$ is an α-finite map.

Consider now the map $\eta \to \rho(\eta)$. Define

$$S(\eta,z+1) \leftrightarrow$$
$$(\exists t)[t\colon z+1 \to H_\eta \ \& \ t \text{ is strictly increasing } \&$$
$$(\exists \sigma)(\forall \tau > \sigma)(\forall x<z)(t(x) \in (\alpha - A^\tau) \ \& \ t(z) = r(\eta))].$$

As before one argues that S in this case is also α-finite, and observe that $\rho(\eta)$ is the least z such that $S(\eta,z+2)$.

We continue with the proof of Theorem 5.21. We will construct an α-r.e. set C which contains α - A but which is disjoint from $R = \{r(\eta) \mid \eta \in K\}$. Now whereas α - A is $\Pi_1(L_\alpha)$, the set R is $\Sigma_2(L_\alpha)$. Elements in R may look sufficiently like elements in α - A to be enumerated into C, and the difficulty of the construction lies in ensuring that this does not happen.

Define $T'(\sigma,\eta,\gamma)$ in the following way. If $\gamma < \rho(\eta)$ or $\rho(\eta) < \gamma$, then

$$T'(\sigma,\eta,\gamma) = \text{the least element in } H_\eta \cap (\sigma - A^\sigma) > \sup_{\delta<\gamma} \{T'(\sigma,\eta,\delta)\}.$$

If $\gamma = \rho(\eta)$, then

$$T'(\sigma,\eta,\gamma) = \text{the least ordinal in } H_\eta \cap ([f'(\sigma,\nu),\sigma) - B^\sigma), \text{ where } k(\nu) = \eta, \text{ greater than } \sup_{\delta<\gamma} \{T'(\sigma,\eta,\delta\}.$$

Clearly T' is an α-recursive function defined for each $\sigma < \alpha$, $\eta \in K$, and $\gamma < \kappa(\eta)$.

We leave the verification of the following result to the reader.

5.23 LEMMA For each $\tau < \sigma$ and $\gamma < \delta$, $T'(\tau,\eta,\gamma) \leq T'(\sigma,\eta,\delta)$. Furthermore, $\lim_{\sigma\to\alpha} T'(\sigma,\eta,\gamma) = T(\eta,\gamma)$ exists for each η, and is the γth element of $H_\eta \cap ((\alpha - A) \cup \{r(\eta)\})$.

Let

$$D_\eta = \{\gamma \mid \rho(\eta) < \gamma < \kappa(\eta) \ \& \ (\exists\sigma)[T'(\sigma,\eta,\gamma) = T(\eta,\rho(\eta))]\}$$

Observe that $T(\eta,\rho(\eta)) = r(\eta)$. Then D is the set of γ's between $\rho(\eta)$ and $\kappa(\eta)$ which approximate to the wrong value $r(\eta)$ at some stage σ. Let D_η^+ be the set of γ's in D_η such that for some stage $\tau, T'(\tau,\eta,\gamma) = T(\eta,\gamma)$ and $T'(\tau,\eta,\rho(\eta)) < T(\eta,\rho(\eta))$, i.e. γ arrives at the correct value under T' before $\rho(\eta)$ does.

5.24 LEMMA The sets D_η and D_η^+ are finite.

Proof. It is sufficient to show that D_η is finite. Suppose that $\{\gamma_n\}_{n<\omega} \subset D_\eta$ is an infinite increasing sequence of ordinals. For each n, let σ_n be the least σ such that $T'(\sigma,\eta,\gamma_n) = T(\eta,\gamma_n) = r(\eta)$. Then $\{\sigma_n\}_{n<\omega}$ is an infinite decreasing sequence of ordinals, which is false.

Let $\kappa = \max \{\kappa(\eta)\}$, for $\eta \in K$. It follows that the set $U = \{(\eta,\gamma) \mid \gamma \in D_\eta\}$ is a tame $\Sigma_2(L_\alpha)$ subset of $K\times\kappa$. Now since the order type of $(\alpha - A)$ is less than $t\sigma 2p(\alpha)$, we have $\kappa < t\sigma 2p(\alpha)$. If we give $K\times\kappa$ the lexicographical ordering, we have by the fact that each D_η is finite the order type of U to be bounded by order type$(K) < \lambda$. This allows one to regard U to be a tame $\Sigma_2(L_\alpha)$ subset of K of order type $< \lambda$. Hence U is α-finite. A similar argument applies to D_η^+ as well.

Hence the maps $\eta \to D_\eta$ and $\eta \to D_\eta^+$ are α-finite. We will need this fact in the definition of the following function:

For each $\gamma \in D_\eta^+$, let $h(\sigma,n,\gamma)$ be the least stage τ such that $T'(\tau,\eta,\rho(\eta)) > T'(\sigma,\eta,\gamma)$. Then h is α-recursive. Let $h_0(\sigma,\eta,\gamma) = \sigma$. By induction, let

$$h_{n+1}(\sigma,\eta,\gamma) = h(h_n(\sigma,\eta,\gamma),\eta,\gamma).$$

Observe that $h(\sigma,\eta,\gamma)$ is defined except when $T(\eta,\gamma)$ is less than $T'(h_{n-1}(\sigma,\eta,\gamma),\eta,\gamma)$. In fact since γ belongs to D_η^+, we have $\rho(\eta) < \gamma$ and so $T(\eta,\rho(\eta)) < T(\eta,\gamma)$. The functions h_n compute the stages σ when $T'(\sigma,\eta,\gamma)$ gives the wrong value for $T(\eta,\gamma)$. Let σ be a stage where $T'(\sigma,\eta,\gamma) = T(\eta,\rho(\eta))$. Let τ be a stage such that $T'(\tau,\eta,\gamma) = T(\eta,\gamma)$ and $T'(\tau,\eta,\rho(\eta)) < T(\eta,\rho(\eta))$. Then $h(\sigma,\eta,\gamma) > \tau$. This implies that $T'(h(\sigma,\eta,\gamma),\eta,\gamma) = T(\eta,\gamma)$. Given n, let σ_n be the least stage σ such that

$$(5.25) \quad T'(h_n(\sigma,\eta,\gamma),\eta,\gamma) = T(\eta,\gamma).$$

By the definition of h, one sees that the set $\{\sigma_n\}$ is a decreasing sequence on n, so that there is a largest n, denoted $n_\eta(\gamma)$, such that (5.25) holds for σ_n. We write $n_\eta(\gamma)$ simply as $n(\gamma)$ if the context is clear. Now if $\sigma < \sigma_{n(\gamma)}$ and if $h_n(\sigma,\eta,\gamma)$ is defined, then $T(\eta,\gamma) > T'(h_n(\sigma,\eta,\gamma),\eta,\gamma)$. The reason for considering the functions h_n and the integer $n(\gamma)$ is that it allows one to pick out elements of $\alpha - A$ in H_η while 'skipping' over $r(\eta)$.

<u>5.26 LEMMA</u> Let $\gamma \in D_\eta^+$. The for all stages σ, if the value $h_{n(\gamma)}(\sigma,\eta,\gamma)$ is defined, then

$$T'(h_{n(\gamma)}(\sigma,\eta,\gamma),\eta,\gamma) \neq T(\eta,\rho(\eta)).$$

Also there is a stage σ such that $T'(h_{n(\gamma)}(\sigma,\eta,\gamma),\eta,\gamma) = T(\eta,\gamma)$.

Proof. If $T'(h_{n(\gamma)}(\sigma,\eta,\gamma),\eta,\gamma) = T(\eta,\rho(\eta))$, then as observed above we have

$$T'(h_{n(\gamma)+1}(\sigma,\eta,\gamma),\eta,\gamma) = T(\eta,\gamma).$$

This of course contradicts the maximality of $n(\gamma)$. On the other hand, if we set $\sigma = \sigma_{n(\gamma)}$, then $T'(h_{n(\gamma)}(\sigma,\eta,\gamma),\eta,\gamma) = T(\eta,\gamma)$.

The reader may verify for himself that $\{(\eta,\gamma,n(\gamma)) \mid \gamma \in D_\eta^+\}$ is tame $\Sigma_2(L_\alpha)$ and bounded below $t\sigma 2p(\alpha)$. Hence $(\eta,\gamma) \to n_\eta(\gamma)$ is an α-finite map.

Now let $E_\eta = D_\eta - D_\eta^+$. Clearly $\eta \to E_\eta$ is α-finite. Let $e_\eta(m)$ denote the mth element of E_η. Set $e_\eta(0)$ to be $\rho(\eta)$. Then

(5.27) For each σ, $T'(\sigma,\eta,e_\eta(0)) < T'(\sigma,\eta,e_\eta(m))$ if $0 < m$.

(5.28) If $\gamma \in E_\eta$ and $T'(\sigma,\eta,\gamma) = T(\eta,\gamma)$, then $T'(\sigma,\eta,\rho(\eta)) = T(\eta,\rho(\eta))$.

We need one more result before showing that B is not a major subset of A. Property (d) in the next lemma is the counterpart of Lemma 5.24 for $\gamma \in E_\eta$.

5.29 LEMMA There is an α-recursive function u' such that

(a) for each stage σ and each $e_\eta(m)$ in E_η, $u'(\sigma,\eta,m)$ is defined;

(b) for each $m < \text{card}(E_\eta)$, $\lim_{\sigma\to\alpha} u'(\sigma,\eta,n) = u(\eta,n)$ exists for all $n < m$;

(c) the graph of u is α-finite;

(d) for all σ, if $u'(\sigma,\eta,m) = u(\eta,m)$, then $T'(\sigma,\eta,en^{(m)}) \neq$

$T(\sigma,\rho(\eta)) = r(\eta)$.

Proof. Let $e_\eta(m)$ be in E_η. Fix σ. Let σ_0 be the least $\tau < \sigma$ such that $T'(\tau,\eta,e_\eta(m)) = T'(\sigma,\eta,e_\eta(m))$. If $T'(\sigma_0,\eta,e_\eta(0)) = T'(\zeta,\eta,e_\eta(m))$ for no $\zeta < \sigma_0$, set $u'(\sigma,\eta,m) = 0$. Otherwise choose the least such ζ and call it ζ_0. Let $u'(\sigma,\eta,m) = u'(\zeta_0,\eta,m)+1$.

Observe that (a) holds by the way u' is defined. Also $u'(\sigma,\eta,m) = u'(\tau,\eta,m)$ if $T'(\sigma,\eta,e_\eta(m)) = T'(\tau,\eta,e_\eta(m))$. This implies that (b) holds. On the other hand, the graph of u is a $\Sigma_2(L_\alpha)$ subset of $K\times\omega\times\omega$. Now $K \subseteq \lambda$, and λ is an α-cardinal. Since $\lambda < \sigma 2p(\alpha)$, we conclude that u is α-finite. This proves (c).

Now let $u'(\sigma,\eta,m) = u(\eta,m)$. Suppose that we have $T'(\sigma,\eta,e_\eta(m)) = x = T(\eta,\rho(\eta))$. Let ζ_0 be the least ζ such that $T(\eta,e_\eta(m)) = T'(\zeta,\eta,e_\eta(m))$. Then $\sigma < \zeta_0$. By (5.28) $T'(\sigma,\eta,\rho(\eta)) = T(\eta,\rho(\eta))$. Since $T'(\zeta_0,\eta,e_\eta(m)) = T(\eta,e_\eta(m))$ we have $u'(\zeta_0,\eta,m) = u(\eta,m)$. Choose ξ_0 to be the least ξ such that $\xi < \sigma$ and $T(\eta,\rho(\eta)) = T'(\xi,\eta,e_\eta(m))$. We get $u'(\sigma,\eta,m) = u'(\xi_0,\eta,m)$ and $T'(\zeta_0,\eta,\rho(\eta)) = T(\eta,\rho(\eta)) = T'(\xi_0,\eta,e_\eta(m))$ so that $u'(\zeta_0,\eta,m) = u'(\xi_0,\eta,m)+1$. This gives $u'(\sigma,\eta,m) \neq u(\eta,m)$, contradicting the choice of σ.

We now define an α-r.e. set C which witnesses the fact that B is not a major subset of A. Recall the definition of the integer $n_\eta(\gamma)$ for $\gamma \in D_\eta^+$. Define C to be the union of the following sets:

(5.30) $\cup\{H_\eta \mid \eta \notin K\}$

(5.31) $\{T'(\sigma,\eta,\gamma) \mid \sigma < \alpha \ \&\ \eta \in K \ \&\ \gamma \notin D_\eta\}$

(5.32) $\{T'(\sigma,\eta,e_\eta(m)) \mid 1 \le m \le \mathrm{card}(E_\eta) \ \&\ u'(\sigma,\eta,m) = u(\eta,m)\}$

(5.33) $\{T'(h_{n_\eta(\gamma)}(\sigma,\eta,\gamma),\eta,\gamma) \mid \gamma \in D_\eta^+\}$.

The set C contains $\alpha - A$. Lemmas 5.24 and 5.29 ensure that $C \cap R = \emptyset$. This completes the proof of Theorem 5.21. Together with

Theorem 5.4, we arrive at the following characterization of the existence of major subsets:

<u>5.34 THEOREM</u> Every non-α-recursive α-r.e. set has a major subset if and only if $\sigma 2cf(\alpha) > \sigma 2p(\alpha)$.

<u>5.35 REMARK</u> Using the results of Chapter 6, one may show as a consequence of Theorem 5.21 that under the same hypothesis, no hyperhypersimple α-r.e. set has a major subset (Chong [1979a]).

<u>5.36 REMARK</u> Theorem 5.21 does not rule out the possibility that major subsets may exist when $\sigma 2cf(\alpha) < \sigma 2p(\alpha)$. It is shown in Chong [1983] that if $\sigma 2cf(\alpha) < \sigma 2p(\alpha) = \alpha$, then every non-$\alpha$-recursive α-r.e. set satisfying certain 'local non-recursiveness' condition has a major subset. A different strategy is required for this. A complete characterization of the existence of major subsets is not known.

CHAPTER 6

HYPERHYPERSIMPLE SETS

We continue the study of the lattic of α-r.e. sets. The notion of a hyperhypersimple set was introduced by Post in his numerous attempts to construct a non-recursive, incomplete r.e. set A by imposing strong conditions on the behavior of the complement of A. Among the possible candidates considered by him were simple sets, hypersimple sets, hyperhypersimple sets, and maximal sets. As subsequent events have shown, none of these notions necessarily implies Turing incompleteness. Nevertheless, Lachlan [1968] has used the class of hyperhypersimple sets fruitfully in the study of the decidability of the elementary theory of $\mathcal{E}(\omega)$. Indeed he showed that this class together with the class of co-finite r.e. sets form a relatively complemented sublattice of $\mathcal{E}(\omega)$, and is therefore decidable according to a result of Ershov. The role that hyperhypersimple sets play in the study of the decidability of the elementary theory of $\mathcal{E}(\alpha)$ should therefore be explored. As a start, one should investigate the problem of existence for these sets. This was carried out in Chong and Lerman [1976]. It is known that no regular constructible cardinals have hyperhypersimple sets, and that if α is a Σ_2 inadmissible constructible cardinal (e.g. $\aleph^L_\omega$), then an α-r.e. set A is hyperhypersimple if and only if its complement is unbounded and has order type less than α. We will prove these results in this chapter.

The reader will note that in contrast to the topics treated earlier, results discussed in this chapter are incomplete. In particular, only special classes of admissible ordinals are considered. One may in fact prove results more general (as in Chong and Lerman [1976]) than what we will present here, at the expense of adding more technical conditions to the hypothesis. There are however two reasons preventing us from doing so. First of all since the problem of existence is not completely solved, it is perhaps more instructive to study two representative cases to see how the key ideas in the construction of these known cases work, and not to be overwhelmed by the additional conditions which have to be imposed in order to prove a more general result. Secondly the class of hyperhypersimple sets should relate directly to the decidability of lattice theory of α-r.e. sets, and in this instance the primary candidates for investigations are $\aleph_\omega^L$ and $\aleph_1^L$, so that from the utilitarian point of view, not very much is missed.

Unless otherwise indicated, α will denote a constructible cardinal for the rest of the chapter.

<u>6.1 DEFINITION</u> An α-r.e. set A is hyperhypersimple (hh-simple) if α - A is not α-finite and the class of α-r.e. sets $B \supseteq A$ forms a boolean algebra under set inclusion.

This definition deviates from the one originally given by Post, where an α-r.e. set A is hh-simple if there is no recursive sequence of pairwise disjoint r.e. sets $\{T_i | i < \omega\}$ such that each T_i intersects α - A non-trivially. This definition implies that every hh-simple set is simple. Lachlan [1968a] however showed the equivalence of these two definitions. Definition 6.1 is preferred since it shows as an immediate consequence the definability in $\mathcal{E}(\alpha)$ of the class of hh-simple sets. Recent developments in classical and higher recursion theory indicate the significant role played by this class in the study of $\mathcal{E}(\alpha)$. It should also be noted that the boolean algebraic approach to the definition of hh-simple sets leads directly and unambiguously to

Definition 6.1 for α-recursion theory, while there are several inequivalent possible generalizations of Post's definition in the general setting, depending, for example, what ordinal γ should be used to replace ω in $\{T_i | i < \omega\}$. Nonetheless, Post's definition turns out to be quite useful in identifying sets which are not hh-simple.

<u>6.2 LEMMA</u> Let A be an α-r.e. set. Let $\{T_i | i < \alpha\}$ be an α-recursive sequence of pairwise disjoint α-r.e. sets such that each T_i intersects $\alpha - A$ non-trivially. Then A is not hh-simple.

Proof. Let T_i^σ be the set of elements enumerated into T_i at stage σ, as given by the simultaneous enumeration. Let

$$B = A \cup \{x | (\exists i)(\exists \sigma)[x \in T_i^\sigma \cap W_i^\sigma]\}.$$

Then $B \supseteq A$ and is α-r.e. We show that $A \cup (\alpha - B)$ is not α-r.e., so that the α-r.e. sets containing A do not form a boolean algebra. Suppose $A \cup (\alpha - B) = W_j$. By hypothesis, there is an $x \in T_j \cap (\alpha - A)$. Then x is in $\alpha - B$ if and only if x is in $A \cup (\alpha - B) = W_j$, if and only if x is in $T_j \cap W_j$, by definition if and only if x is in B. This is a contradiction.

We first consider the case when α is a Σ_2 inadmissible constructible cardinal (e.g. $\aleph_\omega^L$).

<u>6.3 THEOREM</u> Let α be a Σ_2 inadmissible constructible cardinal. An α-r.e. set A is hh-simple if and only if its complement is unbounded in α and has order type less than α.

The proof of Theorem 6.3 is divided into two parts.

<u>6.4 THEOREM</u> Let α be a Σ_2 inadmissible constructible cardinal. If A is α-r.e. and has an unbounded complement of order type

less than α, then A is hh-simple.

Proof. The Σ_2 inadmissibility of α is only used to ensure that α-r.e. sets A with unbounded complement of small order type (i.e. less than α) exist. Let $\beta < \alpha$ be the order type of $\alpha - A$. Let $\{a_i | i < \beta\}$ be an enumeration of $\alpha - A$ in increasing order. If $B \supseteq A$, then $\{i | a_i \notin B\}$ is α-finite. So to show that A is hh-simple, it is sufficient to show that if $K \subseteq \beta$ is an α-finite set, then $A \cup \{a_i | i \in K\}$ is α-r.e.

Let a_i^σ be the ith element in $\alpha - A^\sigma$. We define a function u': $\alpha\times\beta\times\beta \to \omega$ in a way similar to, but simpler than, the one given in Lemma 5.27. This function will 'count' the number of times errors were made in thinking that the ith element of $\alpha - A$ was the jth element of $\alpha - A$, for $i < j$. For each σ, we define only the value of $u'(\sigma,j,i)$ for $j > i$. This is done in the following way: let ζ' be the least ζ such that $a_j^\zeta = a_j^\sigma$. Consider $a_i^{\zeta'}$. If there was no $\xi < \zeta'$ such that $a_i^{\zeta'} = a_j^\xi$, set $u'(\sigma,j,i) = 0$. Otherwise, let ξ' be the least $\xi < \zeta'$ where $a_j^\xi = a_i^{\zeta'}$. Set $u'(\sigma,j,i) = u'(\xi',j,i) + 1$.

Set $u(j,i) = \lim_{\sigma\to\alpha} u'(\sigma,j,i)$ if the limit exists. Then u: $\beta\times\beta \to \omega$ is a function whose graph, being bounded in α, is α-finite. Observe that $u(j,i)$ is in fact defined whenever $j > i$. This follows from the fact that A is regular (since $\alpha = \alpha^*$). Namely, for each j, there is a stage $\sigma(j)$ such that $A|(a_j+1) = A^{\sigma(j)}|(a_j+1)$. Then $u'(\sigma,j,i,) = u'(\sigma(j),j,i)$ for all $\sigma > \sigma(j)$. Furthermore, $u'(\sigma(j),j,i)$ is a finite ordinal by the definition of u' and by the fact that there is no infinite descending sequence of ordinals. We make the following claim:

<u>Claim</u> Fix $j > i$. If $u'(\sigma,j,k) = u(j,k)$ for all $k < j$, then $a_j^\sigma \neq a_i$.

Suppose for the sake of contradiction that $u'(\sigma,j,i) = u(j,i)$ and $a_j^\sigma = a_i$. Choose ζ' to be the least ζ such that $a_j^\zeta = a_j$. Let k be chosen such that $a_i = a_k^{\zeta'}$. Then $k > i$. By the definition of u', we have $u(j,k) = u'(\zeta',j,k)$. Moreover, if ξ' is the least ξ such that $a_j^\xi = a_i$, then $u'(\zeta,j,k) = u'(\xi',j,k) + 1$. It follows that $u'(\sigma,j,k) = u'(\xi',j,k) \neq u'(\zeta',j,k) = u'(j,k)$, contradicting the assumption. This proves the claim.

Now let K be an α-finite subset of β. If $X = \{a_i | i \in K\}$ is bounded in α, then it is α-finite and so clearly $A \cup X$ is α-r.e. Suppose that X is not bounded in α. Let

$$B = A \cup \{x | (\exists\sigma)(\exists j)[j \in K \ \& \ a_j^\sigma = x \ \& \ u'(\sigma,j,k) = u(j,k) \text{ for } k < j]\}.$$

By the claim, each x enumerated in B at no time equals a_i for $i \notin K$. This means that if $j \in K$, and $x = a_j^\sigma$ and $u'(\sigma,j,k) = u(j,k)$ for all $k < j$, then the claim assures that $x \neq a_i$ for all $i < j$. Clearly one cannot have $x = a_i$ for some $i > j$. It follows that if $x \neq a_j$, then $x \in A$. We may therefore conclude that $B = A \cup X$. This proves Theorem 6.4.

<u>6.5 THEOREM</u> Let α be a Σ_2 inadmissible cardinal. If A is α-r.e. and has complement of order type α, then A is not hh-simple.

Proof. Let $\{H_\eta | \eta < \sigma2cf(\alpha)\}$ be the simultaneous α-r.e. sequence as given in Lemma 5.18. Thus the H_η's are pairwise disjoint, $\cup \{H_\eta | \eta < \lambda\}$ is α-finite for each $\lambda < \sigma2cf(\alpha)$, and every x belongs to some H_η. Let $f: \sigma2cf(\alpha) \to \alpha$ be a $\Sigma_2(L_\alpha)$ cofinal function approximated by the α-recursive function $f': \alpha\times\sigma2cf(\alpha) \to \alpha$. Since α is Σ_2 inadmissible, it is a limit of constructible cardinals and so we may choose $f(\nu)$ to be a constructible cardinal for each $\nu < \sigma2cf(\alpha)$ (since being an α-cardinal is a $\Pi_1(L_\alpha)$-relation). Furthermore, we may assume that f is strictly increasing and that $f'(\sigma,\nu) > f'(\tau,\eta)$ whenever $\sigma > \tau$

and $\nu > \eta$. We will construct an α-r.e. sequence $\{T_i | i < \alpha\}$ as in Lemma 6.2 to show that A is not hh-simple.

For each $\nu < \sigma 2cf(\alpha)$, let $u(\nu)$ be the least η such that $|H_\eta \cap (\alpha - A)|^L > f(\nu+1)$, where $|X|^L$ denotes the (constructible) cardinality of a set X. The ordinal $u(\nu)$ is defined for each ν less than $\sigma 2cf(\alpha)$ since the order type of $\alpha - A$ is α. Then $u: \sigma 2cf(\alpha) \to \sigma 2cf(\alpha)$ is an α-finite map. We divide the ordinal α into $\sigma 2cf(\alpha)$ many blocks. For each i such that $f(\nu) \leq i < f(\nu+1)$ (say that i is in the νth block), we define an α-r.e. set T_i with the intention of putting an $x \in (H_{u(\nu)} \cap (\alpha - A))$ into T_i. Steps will be taken to ensure that $\{T_i | i < \alpha\}$ is an α-recursive sequence of pairwise disjoint α-r.e. sets. Since there are only $f(\nu+1)$ many i's between $f(\nu)$ and $f(\nu+1)$, while by the definition of the function u there are at least $f(\nu+2)$ many (in cardinality) elements of $\alpha - A$ in $H_{u(\nu)}$, there is plenty of opportunity to assign one such x to each T_i for i in the νth block. Lemma 7.2 will then show that A is not hh-simple. We proceed with the definition of $\{T_i | i < \alpha\}$:

We say that i is in the νth block at stage σ if $f'(\sigma,\nu) \leq i < f'(\sigma,\nu+1)$. At stage σ, look for the least i (called $i(\sigma)$) such that i is in the νth block (at σ) for some ν and there is no x in both $T_i^{<\sigma}$ and $H_{u(\nu)} \cap (\alpha - A^{<\sigma})$. Compute σ steps to obtain an x (if exists) in $H_{u(\nu)} \cap (\alpha - A^\sigma)$ which has not already been enumerated into T_j for any j. Let $T^\sigma_{i(\sigma)} = T^{<\sigma}_{i(\sigma)} \cup \{x\}$. Let $T_j^\sigma = T_j^{<\sigma}$ for $j \neq i(\sigma)$. This completes the construction at stage σ.

Clearly $\{T_i | i < \alpha\}$ is a simultaneous α-r.e. sequence of pairwise disjoint α-r.e. sets. We show that $T_i \cap (\alpha - A)$ is non-empty for each i. Suppose that i is in the νth block. By the regularity of A (since $\alpha = \alpha^*$), and the fact that $\nu+1 < \sigma 2cf(\alpha)$, there is a stage σ where $A|\sup(H_{u(\nu)}) = A^\sigma|\sup(H_{u(\nu)})$ and $f'(\tau,j) = f(j)$ for all $\tau > \sigma$ and j in the νth block. At stage σ, there are at least $f(\nu+2)$ many elements in $H_{u(\nu)} \cap (\alpha - A^{\sigma})$, and these are elements in $H_{u(\nu)}$ which belong to $\alpha - A$. Moreover, these elements are reserved exclusively for the use of ordinals j in the νth block, and there are only $f(\nu+1)$ many such j's. Thus if before stage σ there was no x in $H_{u(\nu)}$ enumerated into T_i which remains to appear like an element of $\alpha - A$, then it is certainly possible to choose an x in $H_{u(\nu)} \cap (\alpha - A^{\sigma})$ not already in any $T_j^{<\sigma}$ and put it into T_i. Furthermore by the choice of the stage σ, this is a genuine member of $\alpha - A$. This implies that the sequence $\{T_i | i < \alpha\}$ is a simultaneous α-r.e. sequence of pairwise disjoint α-r.e. sets, each intersecting $\alpha - A$ non-trivially. By Lemma 6.2, A is not hh-simple. This proves Theorem 6.5.

Theorem 6.3 is now an immediate consequence of Theorems 6.4 and 6.5.

It is worth noting that one may extend Theorem 6.3 to all Σ_2 inadmissible ordinals which are not Σ_2 projectible. This is made precise in the following way.

<u>6.6 LEMMA</u> Suppose that $\alpha = \sigma 2p(\alpha) > \sigma 2cf(\alpha)$. Then α is a limit of α-cardinals.

Proof. Assume to the contrary that κ is the largest α-cardinal. Let $f: \sigma 2cf(\alpha) \to \alpha$ be a $\Sigma_2(L_\alpha)$ cofinal function. Let $u(\nu)$ be an α-finite injection of $f(\nu+1)$ into κ. For each $i < \alpha$, if i is in the νth block (as defined in the proof of Theorem 6.5), let $F(i) = \langle u(\nu)(i), i\rangle$. Then $F: \alpha \to \kappa\times\kappa$ is an injection. It is furthermore $\Sigma_2(L_\alpha)$. By Theorem 1.28, we have $\sigma 2p(\alpha) < \alpha$ and this is a contradiction.

6.7 THEOREM Suppose that $\alpha = \sigma 2p(\alpha) > \sigma 2cf(\alpha)$. An α-r.e. set A is hh-simple if and only if its complement is unbounded and has order type less than α.

Proof. Retrace the steps of the proof given in Theorem 6.4 and 6.5. First of all, let A be α-r.e. whose complement has order type $\beta < \alpha$. Then the map u given in 6.4 is $\Sigma_2(L_\alpha)$ and its graph is a subset of $\beta \times \beta \times \omega$. Since $\sigma 2p(\alpha) = \alpha$, the graph of u is α-finite. The rest of the argument follows and so A is hh-simple. Next, suppose that A has complement of order type α. In this case the map u defined in 6.5 is also $\Sigma_2(L_\alpha)$ (recall that the notion of an α-cardinal is $\Pi_1(L_\alpha)$), and its graph is a subset of $\sigma 2cf(\alpha) \times \sigma 2cf(\alpha)$, hence α-finite. Since by Lemma 6.6. we have α to be a limit of α-cardinals, the argument of Theorem 6.5 can be mimicked to show that A is not hh-simple. We leave the details to the reader.

We now show that there is a large class of admissible ordinals (e.g. $\aleph_1^L$) for which hh-simple sets do not exists.

6.8 THEOREM Let α be a regular constructible cardinal. Then there is no hh-simple α-r.e. set.

Proof. Assume on the contrary that H is an hh-simple α-r.e. set. Then the complement of H has order type α. Write the complement as $\{h_i \mid i < \alpha\}$ (in increasing order), and let h_i^σ be the ith element (in magnitude) of $\alpha - H^\sigma$. Clearly if $x = h_i^\tau = h_j^\sigma$ for $\sigma > \tau$, then $i \geq j$. It follows that for each ordinal x, $\{i \mid x = h_i^\sigma$ for some $\sigma\}$ is finite, for otherwise there will be an infinite descending sequence or ordinals. For each natural number n, let C_n be the set of ordinals x such that $\{i \mid x = h_i^\sigma$ for some $\sigma\}$ has cardinality at least n. Then

(6.9) C_n is α-r.e.

(6.10) $\bigcup \{C_n | n < \omega\} = \alpha$.

(6.11) Every x fails to be in C_n for all sufficiently large n.

(6.12) If $x \in C_n$, then $x \in C_m$ for $m \leq n$.

We first prove a lemma. Recall that $[\nu,\delta) = \{x | \nu \leq x < \delta\}$.

<u>6.13 LEMMA</u> Fix $\delta < \alpha$. There is no α-recursive relation R such that for all n and $x \in (\alpha - H) \cap [\delta,\alpha)$,

$$x \in C_n \leftrightarrow (x,n) \in R.$$

Proof. Suppose such an α-recursive relation exists. Observe first of all that there is an n such that

$$(6.14)\ C(n) = C_n \cap (\alpha - H) \cap (\alpha - (\bigcup C_m, m > n)$$

is unbounded. If not, let c(n) be an upper bound for ordinals in the set defined by (6.14). Since α has uncountable constructible cofinality, there is a uniform upper bound ζ for the set $\{c(n) | n < \omega\}$. This means that no ordinal greater than ζ is in the set C(n) defined by (6.14) for any n. Since the complement of H has order type α, there is an $h_i > \zeta$. By (6.10) h_i belongs to C_n for some n. By (6.11) we may choose the largest such n. But then h is an element of C(n), contradiction.

Thus let n^* be the least natural number n such that C(n) is unbounded in α. Let ξ be chosen which bounds all ordinals in C(n) for $n < n^*$. Let

$$D = \{x | x > \xi \ \&\ (x,n^*) \in R \ \&\ (x,n^*+1) \notin R\}.$$

D is an α-r.e. set (since by (6.9) C_{n^*} is). For $x \in D$, we say that x has jumped n^* times at stage σ if $|\{i|(\exists\tau<\sigma)[x = h_i^\tau]\}| = n^*$. This implies that for all $\nu > \sigma$, if $x = h_j^\nu$ then $x = h_j^\sigma$ (so that either x ceases to be an element of $\alpha - H^\nu$ or its order (in magnitude) in $(\alpha - H^\nu)$ is the same as its order in $(\alpha - H^\sigma)$). We now construct a simultaneous α-r.e. sequence $\{T_i|i < \alpha\}$ of pairwise disjoint α-r.e. sets such that each T_i intersects $(\alpha - H)$ nontrivially.

At stage σ, let $i(\sigma)$ be the least i such that there is no $x \in T_i^{<\sigma} \cap (\alpha - H^{<\sigma})$. Go to the least stage $\sigma' > \sigma$ for which there is an $x \in D$ not in any $T_j^{<\sigma}$, $j < \alpha$, such that x has jumped n^* times at stage σ' and $x \in (\alpha - H^{<\sigma})$. Let $T_{i(\sigma)}^\sigma = T_{i(\sigma)}^{<\sigma} \cup \{x\}$ and let $T_j^\sigma = T_j^{<\sigma}$ for $j \neq i(\sigma)$. This completes the construction of the sequence $\{T_i|i < \alpha\}$ at stage σ.

We claim that T_i intersects $\alpha - H$ non-trivially. Otherwise, pick the least i, denoted i^*, such that $T_i \cap (\alpha - H) = \emptyset$. By our construction there is a least stage σ where for all $j < i^*$, T_j^σ already contains an element of $D \cap (\alpha - H)$. Denote this stage by σ^*. By the definition of D, there are unboundedly many elements of $\alpha - H$ in D, and so by the construction (which always looks for such elements) there is a stage $\nu > \sigma^*$ where an $x \in (\alpha - H)$ was chosen to enter some T_j, $j > i^*$. Suppose that ν is the least such stage. Then $x = h_s$ is the least such element in $\alpha - H$ being chosen to enter some T_j. Furthermore since $x \notin T_{i^*}$, there must be elements y which enter T_i's, for $i^* \leq i < j$ at stages between σ^* and ν. Every such y fails to be in $\alpha - H$ (since x is the least element to be found in $\alpha - H$). At stage ν, clearly there is a $y(i) \in T \cap (\alpha - H^{<\nu}) \cap D$, for each i such that $i^* \leq i < j$ (otherwise x would enter some T_i where $i < j$). There is however a stage $\rho > \nu$ such that $y(i) \in H$ for all i, $i^* \leq i < j$. Indeed since we are only interested in ordinals above ξ, every such ordinal below x and greater than $x(i)$ (the element in $T_i \cap (\alpha - H)$) for $i < i^*$) is in H. This means that at stage ρ, $x = h_t^\rho$ and $t < s$. It follows that x has in fact 'jumped' one more time on its order (from h_s^{ν} to h_t^{p}), and so $(x,n^*+1) \in R$, contradicting the fact that x is in D. Thus each T_i intersects $(\alpha - H)$ non-trivially. This, however, contradicts the hh-simplicity of H by Lemma 6.2.

We now construct a simultaneous α-r.e. sequence of α-r.e. sets $D(n,\nu)$, where $n < \omega$ and $\nu < \alpha$, such that for each n, $D(n,\nu) \subset C_n$. The idea is to attempt to make each $D(n,\nu)$, for some fixed n, intersect $\alpha - H$ non-trivially. Since H is hh-simple, this attempt will fail. But then one shows that in this case there is an α-recursive relation R as described in Lemma 6.13, which is a contradiction.

At stage σ for each n, look for the least ν such that there is no $x \in D(n,\nu)^{<\sigma} \cap (\alpha - H^{<\sigma}) \cap C_n^{<\sigma}$, and such that there is an x in $(\alpha - H^{\sigma}) \cap C_n^{\sigma}$ greater than every element y of $\alpha - H^{\sigma}$ in $D(n,\eta)$ where $\eta < \nu$ and x is not in any $D(n,\zeta)$ thus far. Let $D(n,\nu)^{\sigma} = D(n,\nu)^{<\sigma} \cup \{x\}$. Let $D(n,\eta)^{\sigma} = D(n,\eta)^{<\sigma}$ for $\eta \neq \nu$. As usual let $D(n,\eta) = \bigcup D(n,\eta)^{\sigma}$ for each n and $\eta < \alpha$.

Now by the hh-simplicity of H, for each n there is a ν such that $D(n,\nu) \cap (\alpha - H) = \emptyset$. Let $\nu(n)$ be the least such ν. Then since α has uncountable constructible cofinality, there is a uniform bound for $\{\nu(n) \mid n < \omega\}$. It follows that the graph of the map $n \to \nu(n)$ is an α-finite set K. Now since $D(n,\nu(n))$ does not contain an element of $\alpha - H$, unboundedly many attempts will be made on the set $D(n,\nu(n))$ to accept a (potential) member of $\alpha - H$. Since each of these attempts failed, $D(n,\nu(n))$ must be unbounded in α. On the other hand, for each $\nu < \nu(n)$, $D(n,\nu) \cap (\alpha - H)$ is non-empty, and by construction contains precisely one member $x(n,\nu)$ of $\alpha - H$. Moreover, once $x(n,\nu)$ is enumerated into $D(n,\nu)$ (at stage σ say), $D(n,\nu)$ will stop growing. It follows that for each n and each $\nu < \nu(n)$, there is a stage $\sigma(n,\nu)$

such that $D(n,\nu)^{\rho} = D(n,\nu)$ for all $\rho > \sigma(n,\nu)$. Since α is a regular constructible cardinal, $\{\sigma(n,\nu) \mid \nu < \nu(n)\}$ is bounded below α for each n. Indeed the regularity of α implies that there is a uniform bound σ for $\{\sigma(n,\nu) \mid n < \omega$ and $\nu < \nu(n)\}$. Assume that σ_0 is the least such σ. Thus at stage σ_0 each $D(n,\nu)^{\sigma_0}$, $n < \omega$ and $\nu < \nu(n)$, contains an element of $\alpha - H$. Also, there is a uniform bound x^* for $x(n,\nu)$ where $n < \omega$ and $\nu < \nu(n)$. Define R in the following way:

$$(x,n) \in R \quad \leftrightarrow \quad x \in C_n^{\sigma} \text{ where } \sigma > \sigma_0 \text{ is the least stage where there is a } y > x \text{ and } y \in C_n^{\sigma} \text{ and } y \text{ is enumerated into } D(n,\nu(n))^{\sigma}.$$

Since $D(n,\nu(n))$ is unbounded for each n, R is clearly α-recursive. We claim that for $x > x^*$ and $x \in \alpha - H$, $(x,n) \in R$ if and only if $x \in C_n$. Choose such an x. Obviously by construction if $(x,n) \in R$ then $x \in C_n$. Suppose now $x \in C_n$. Assume that at the least stage σ where x is enumerated in C_n^{σ}, there is a $y > x$ in $D(n,\nu(n))^{\sigma}$ which appears like an element of $\alpha - H$. Then x is barred from entering any $D(n,\nu), \nu > \nu(n)$, for as long as y continues to appear like an element of $\alpha - H$ (since any z that is enumerated during this period into $D(n,\nu)$, $\nu > \nu(n)$, must be greater than y by construction). This continues to hold until y gets enumerated into H (which must happen since by assumption $D(n,\nu(n)) \cap (\alpha - H) = \emptyset$). But then at this stage (called τ) x is an ordinal which satisfies all the conditions that are required for putting it into $D(n,\nu(n))$. This will imply that $D(n,\nu(n)) \cap (\alpha - H) \neq \emptyset$, a contradiction. We conclude that if an element y of C_n is enumerated into $D(n,\nu(n))$ at stage σ, then all x in $C_n^{\sigma} \cap (\alpha - H)$ such that $x^* < x < y$ are already in C_n. Thus $(x,n) \in R$.

We now have an α-recursive relation R which satisfies the condition prohibited by Lemma 6.13. This is a contradiction and so we conclude that H is not hh-simple. This proves Theorem 6.8.

Remark The above construction may be generalized to show that no Σ_2 admissible ordinal α such that $t\sigma 2p(\alpha) = \alpha$ and $S_3(L_\alpha)$ cofinality $> \omega$ has an hh-simple set (Chong and Lerman [1976]).

CHAPTER 7

MINIMAL PAIRS

A priority construction whose complexity falls somewhere between α-finite injury (as in the proof of the Splitting Theorem) and full α-infinite injury (as in the proof of the Density Theorem to be discussed in the next chapter) is that of minimal pairs. Two α-r.e. sets A and B are said to form a minimal pair if every α-r.e set C which is α-recursive in both A and B is α-recursive. The existence of minimal pairs for ω-r.e. sets were shown independently by Lachlan [1966] and Yates [1966]. Their constructions involved some infinite priority argument, and the key idea has since been considerably clarified by Lerman [1972] (via the notion of a pinball machine) and Soare [1981]. In α-recursion theory, the problem of the existence of a minimal pair for every admissible ordinal remains open, despite the efforts of various people. Sukonick [1969] first constructed a minimal pair of ω_1^{CK}-r.e. set. Lerman and Sacks [1972] obtained a minimal pair of α-r.e. sets whenever α is what they termed 'non-refractory'. Shore [1978] introduced simplifications to the construction of minimal pairs and used these ideas to solve a case left open in Lerman and Sacks [1972]. Maass [1977] also considered the problem when α is a limit of α-stable ordinals. We illustrate Shore's technique by proving the existence of minimal pairs for a wide class of admissible ordinals, including those which are Σ_2 admissible.

7.1 THEOREM If α is one of the following ordinals, then minimal pairs exist:

(a) $\sigma 2p(\alpha) = \alpha$

(b) $\sigma 2cf(\alpha) = \alpha$.

We will construct two α-r.e. α-regular sets A_0 and A_1 satisfying the following two types of requirements:

$(R_e(A_i))$: If W_e is unbounded in α, then $A_i \cap W_e \neq \emptyset$.

(S_e) If $\{e\}^{A_0} = \{e\}^{A_1} = B$, then B is α-recursive.

Requirement $R_e(A_i)$ ensures that the complement of A_i is not α-r.e., so that A_i is not α-recursive. The way to satisfy this requirement is to choose an $x \in W_e^\sigma$ at stage σ and put it in A_i^σ. This by itself does not cause any serious problem, except when the restraints imposed by requirements of type S are taken into consideration. The restraints introduced by S_e are essentially of the following form: Suppose that at stage σ we have

$$(7.2)\quad \{e\}_\sigma^{A_0^\sigma} | \gamma = \{e\}_\sigma^{A_1^\sigma} | \gamma$$

for some γ and there is a requirement $R_d(A_i)$ which meets certain prescribed conditions (to be specified later) and an ordinal x is available to satisfy $R_d(A_i)$ (by putting x in A_i). Doing this may however destroy some computations associated with requirements of type S. To strike a compromise between satisfying $R_d(A_i)$ and S_e, we will allow only the computation on at most one side of equation (7.2) to be injured, retaining the other side of the computation in (7.2). This puts a restraint on the ordinals allowed to enter A_i at stage σ. The computation on the other side of (7.2) that is not injured at stage σ will be preserved, until a later stage $\tau > \sigma$ when (7.2) again becomes valid (with σ replaced by τ). If such a τ does not exist, then one obtains by the regularity of A_i a $y < \gamma$ such that

$$(7.3)\quad \lim_{\sigma\to\alpha} \{e\}_\sigma^{A_0^\sigma}(y) = \{e\}^{A_0}(y) \neq \{e\}^{A_1}(y) =$$

$$\lim_{\sigma\to\alpha} \{e\}_{\sigma}^{A_1^{\sigma}}(y).$$

This implies that the hypothesis of S_e does not hold, and so S_e is trivially satisfied. On the other hand, if the hypothesis of S_e holds, then one may argue (modulo some details) that since at any stage only at most one side of (7.2) is injured (for each γ), the computation for $y < \alpha$ that is not injured (say $\{e\}_{\sigma}^{A_0^{\sigma}}(y)$) will be correct, so that one has in fact $\{e\}_{\sigma}^{A_0^{\sigma}}(y) = \{e\}^{A_0}(y)$. This gives an effective algorithm for computing $\{e\}^{A_0}$. Returning to requirement $R_d(A_i)$, our strategy for satisfying it is to put an x in W_d above all the restraints imposed by requirements S_e into A_i at stage σ, provided of course that $R_d(A_i)$ is the requirement of highest priority at this stage. There are however further complications. Consider the situation where there is an $x \in W_d^{\sigma}$ which is a candidate for satisfying $R_d(A_i)$ except that there is a restraint imposed by S_e preventing it from entering A_i at stage σ. At some stage $\tau > \sigma$ the restraint imposed by S_e is removed and it would seem that x could now enter A_i except that at this point it is prevented to do so because of another restraint imposed by some requirement $S_{e'}$. The situation may persist through the stages less than α, so that the ordinal x never enters A_i, and therefore $R_d(A_i)$ is never satisfied.

Shore [1978] obtained a solution to the problem by exploiting the fact that when $\alpha > \omega$, there exist unboundedly many limit ordinals below α. We let the requirements of type $R_e(A_i)$ be given a priority list of length $\sigma 2p(\alpha)$. Then every $R_e(A_i)$ is associated with an ordinal less than $\sigma 2p(\alpha)$ (its priority) and for every proper initial segment δ of $\sigma 2p(\alpha)$ there is a stage where every $R_e(A_i)$ of priority higher than δ is accorded its correct value thereafter. The idea now is to devise a construction so that for each γ and e, $\{e\}^{A_0}|\gamma$ is injured only finitely many times after the priorities on an appropriate initial segment of requirements have settled down. This is achieved by stipulating that if the computation $\{e\}_\sigma^{A_0^\sigma}|\gamma$ is injured by a requirement $R_d(A_i)$ at stage τ, then no requirement $R_{d'}(A_j)$ of lower priority than $R_d(A_i)$ at stage $\rho > \tau$ may injure it at a stage ρ. Thus if $R_d(A_i)$ has priority δ and all requirements of priority higher than or equal to δ attain their correct values at stage τ and thereafter, then there are at most finitely many stages σ greater than τ where the computation $\{e\}_\sigma^{A_0^\sigma}|\gamma$ gets injured. This is true for the computation associated with the restraint imposed by requirement $S_{e'}$ (say $\{e'\}_\nu^{A_1^\nu}|\gamma'$ is to be protected) as well. As a consequence of this, if (7.2) holds for e and e' respectively on unboundedly many σ and γ, then there is a stage $\rho > \tau$ where it holds for both e and e' simultaneously. This is also true if we consider an α-finite set of e's instead of only e and e' (Lemma 7.6). At stage ρ it is then possible to put $x \in W_d$ enumerated at stage σ into A_i. Requirement $R_d(A_i)$ is then satisfied.

We now proceed with the proof of Theorem 7.1 (a). The length of the priority listing in this case is α. $R_d(Ai)$ has higher priority than $R_{d'}(A_j)$ if $d < d'$. If $d = d'$ then $R_d(A_0)$ has higher priority than $R_d(A_1)$. Also $R_d(A_i)$ has higher priority than S_e if $d < e$. If $d = e$, then S_e has higher priority than $R_d(A_i)$. And finally, S_e has higher priority than $S_{e'}$ if $e < e'$. As usual, the actual ordering given to the requirements is not important. What is important is the length of the priority listing. Given e and σ, the <u>length of agreement</u> γ for the eth reduction procedure at stage σ is the least x where $\{e\}_\sigma^{A_0^S}(x) \neq \{e\}_\sigma^{A_1^\sigma}(x)$. Denote this ordinal by $L(\sigma,e)$. The <u>maximum length of agreement</u> by stage σ is $\sup\{L(\nu,e)|\nu < \sigma\}$. This is denoted $M(\sigma,e)$. The restraint function $R(\sigma,e)$ is defined to be 0 if $L(\sigma,e) = M(\sigma,e)$ (i.e. if the maximum length of agreement is attained at the current stage). Otherwise, it is the least ordinal greater than the supremum of the stages $\nu < \sigma$ where $L(\nu,e) = M(\sigma,e)$ (recall our convention that at stage σ only σ-many steps of computations are carried out, so that $L(\nu,e) < \sigma$ for each $\nu < \sigma$). Restraints are set only by requirements S_e. Let $R^*(\sigma,e)$ be the supremum of $R(\sigma,e')$, where $S_{e'}$ has higher priority than $R_e(A_i)$. The restraints $R^*(\sigma,e)$ are also known as 'water marks' (a term coined by Soare) measuring the level below which no computations obtained to satisfy a higher priority $S_{e'}$ is allowed to be injured for the purpose of satisfying $R_e(A_i)$. The water marks may rise and fall, and an ordinal x which is put into A_i at stage σ to satisfy $R_d(A_i)$, $d > e$, has to be greater than $R^*(\sigma,e)$ in the first place.

Let $A_i^0 = \emptyset$. Let $A_i^{<\sigma} = \bigcup A_i^{\tau}$. The construction is as follows: At stage σ, find the $R_e(A_i)$ with the highest priority such that

(7.4) there is an $x \in W_e^{\sigma} - A_i^{<\sigma}$ with $x > R^*(e,\sigma)$;

(7.5) if $S_{e'}$ has higher priority than $R_e(A_i)$ and there is a γ such that putting x into A_i^{σ} would injure the computation of $\{e'\}^{A_i^{<\sigma}}|\gamma$, then $R_e(A_i)$ is the requirement of highest priority to do so by stage σ.

This completes the construction at stage σ.

7.6 LEMMA Let K be an α-finite set such that for each e' in K, $\{\sigma | R(\sigma,e') = 0\}$ is unbounded in α. Then $\bigcap \{\sigma | R(\sigma,e') = 0\}$, $e' \in K$, is unbounded in α.

Proof. Claim For each $e' \in K$, $\{\sigma | R(\sigma,e') = 0\}$ is also closed (in the set-theoretic sense).

Suppose that λ is a limit ordinal and there is a sequence of ordinals σ cofinal in λ such that $R(\sigma,e') = 0$. For each such σ, by construction we have $L(\sigma,e') = M(\sigma,e')$. But then by (7.5) there are only finitely many stages τ between σ and λ where $L(\tau,e') < L(\sigma,e')$. This implies that at stage λ we have $L(\lambda,e') = M(\lambda,e')$ so that $R(\lambda,e') = 0$. Hence $\{s \mid R(\sigma,e') = 0\}$ is closed and unbounded. It is moreover α-recursive. Since intersection of closed and unbounded (α-recursive) sets are closed and unbounded, we have Lemma 7.6.

7.7 LEMMA For every e and i, $R_e(A_i)$ is satisfied.

Proof. Fix an e and an i. Suppose that W_e is unbounded in α. Let $K_0 = \{d \mid R_d(A_j)$ has higher priority than $R_e(A_i)$ and $W_d \cap A_j \neq 0\}$. Then every $R_d(A_j)$ with d in K_0 is satisfied at some stage. Furthermore, K_0 is an α-r.e. subset of the ordinal e and is therefore α-finite (since $\alpha^* = \alpha$ in this case). It follows by admissibility that there is a stage $\sigma_0 > e$ such that every $R_d(A_j)$ with d in K_0 is satisfied. Let $\sigma_1 > \sigma_0$ be chosen such that each computation of the form $\{e\}_\tau^{A_i^\tau} \mid \gamma$ which is injured by an $R_{e'}(A_j)$ of higher priority has been reinstated (if it ever does). By (7.5) from stage σ_1 onwards, no computation of the type $\{e\}_\tau^{A_i^\tau} \mid \gamma$ ever gets injured again.

Claim There is a Σ_2 admissible ordinal $\sigma_2 > \sigma_1$ such that $\sigma_2 < \alpha$.

It is enough to assume that α is not Σ_2 admissible. Since we have $\sigma 2p(\alpha) = \alpha$, by Lemma 6.6 we must have α to be the limit of α-cardinals. Choose a regular α-cardinal σ_2 greater than σ_1. Then σ_2 is Σ_2 admissible (otherwise the $\Sigma_2(L_{\sigma_2})$ cofinality map, being α-finite, would give a contradiction to the regularity of σ_2).

Let $\sigma_2 > \sigma_1$ be chosen to be Σ_2 admissible such that $W_e^{\sigma_2}$ is unbounded in σ_2. Such an ordinal exists since every α-cardinal is α-stable (as in the proof of Corollary 3.4). The fact that W_e is unbounded in α then ensures that it is unbounded in σ_2 as well. Now let K_1 be the set of all $S_{e'}$ of higher priority than $R_e(A_i)$ such that $R(e',\sigma) = 0$ for unboundedly many σ in σ_2. Let K_2 be $e - K_1$. Both of these sets are α-finite, hence σ_2-finite. For each e' in K_2, define a function G so that $G(e') = \sup\{\sigma | L(\sigma,e') = M(\sigma,e')\}$. Then by Σ_2 admissibility there is a uniform bound $\sigma_3 < \sigma_2$ on all of the $G(e')$'s. Now choose $x \in W_e^{\sigma_2}$ greater than σ_3. By Lemma 7.6 there is a stage $\sigma > \sigma_3$ where $R(\sigma,e') = 0$ for all e' in K_1. At this stage one would put x in A_i to satisfy $R_e(A_i)$. This proves Lemma 7.7.

<u>7.8 LEMMA</u> If $\{e\}^{A_0} = \{e\}^{A_1} = C$, then C is α-recursive.

Proof. Fix γ.

<u>Claim</u> There is a stage σ such that $L(\sigma,e) \geq \gamma$.

To see this, observe that there is a σ such that $M(\sigma,e) = L(\sigma,e) \geq \gamma$. If this is false, choose the least γ which is a counterexample. Then $M(\sigma,e) < \gamma$ for all σ. Suppose $\gamma = \delta + 1$. Then there is a σ_0 where $M(\sigma_0,e) = L(\sigma_0,e) \geq \delta$. Since we have the equality $\{e\}^{A_0} = \{e\}^{A_1}$, there is a stage $\sigma_1 > \sigma_0$ where $\{e\}_{\sigma_1}^{A_0^{\sigma_1}}(\delta+1)$ is equal to $\{e\}_{\sigma_1}^{A_1^{\sigma_1}}(\delta+1)$, so that $M(\sigma_1,e) = L(\sigma_1,e) = \gamma$, contradicting our choice of γ. If on the other hand γ is a limit ordinal, then for each $\gamma' < \gamma$, there is a $\sigma_{\gamma'}$ where $M(\sigma_{\gamma'},e) = L(\sigma_{\gamma'},e) \geq \gamma'$. This is a stage where $R(\sigma_{\gamma'},e) = 0$. By the Claim in Lemma 7.6, $R(\lambda,e) = 0$, where $\lambda = \sup \sigma_{\gamma'}$, for $\gamma' < \gamma$. We then have $M(\lambda,e) = L(\lambda,e) \geq \gamma$, a contradiction.

Now the construction ensures that only one of the two computations $\{e\}^{A_0}|\gamma$ or $\{e\}^{A_1}|\gamma$ is ever injured at any stage, and this gets reinstated at a later stage. This implies that at stage σ, (as in the claim) the computation obtained for the side that is not injured, say $\{e\}_{\sigma}^{A_0^{\sigma}}|\gamma$, is correct (i.e. equal to $\{e\}^{A_0}|\gamma$). Hence C is C is α-recursive.

Lemmas 7.7 and 7.8 together prove Theorem 7.1 (a).

We now consider 7.1 (b), the case when $\sigma 2cf(\alpha) = \alpha$. In the situation when $\sigma 2p(\alpha) < \alpha$, the natural list of priority ordering is no longer sufficient. This is especially apparent when one attempts to prove for example the analogue of Lemma 7.7, where it is required to show that certain subset of a proper initial segment of ordinals is α-finite. To be able to conclude this one needs the fact that the subset is $\Sigma_2(L_\alpha)$ definable, and bounded below $\sigma 2p(\alpha)$. The most reasonable choice of priority listing for this purpose is then $\sigma 2p(\alpha)$. Thus let $g: \sigma 2p(\alpha) \to \alpha$ be a $\Sigma_2(L_\alpha)$ map from a subset of $\sigma 2p(\alpha)$ onto α. This shorter list of requirements by itself is not enough. One needs also the assumption that $\sigma 2p(\alpha) < \sigma 2cf(\alpha)$ in the consideration of requirements of type S_e. We therefore introduce two lists of requirements: $R_{e'}(A_j)$ has higher priority than $R_e(A_i)$ if the least β' such that $g(\beta') = e'$ is less than the least β such that $g(\beta) = e$. $S_{e'}$ has higher priority than S_e if the same situation holds. As before if $e' = e$, then $R_{e'}(A_0)$ will have higher priority than $R_e(A_1)$. We set $S_{e'}$ to have higher priority than $R_e(A_i)$ if the least β' such that $g(\beta') = e'$ is less than or equal to the least β such that $g(\beta) = e$ (and vice versa if $\beta < \beta'$). Let g' be α-recursive approximations of g. Define 'higher priority at stage σ' in a manner analogous to the above, replacing g by $g'(\sigma,.)$. Define the functions $L(\sigma,e)$ and $M(\sigma,e)$ as before. For the restraint function $R(\sigma,.)$, we now modify it so that it is defined on $\sigma < \alpha$ and $\beta < \sigma 2p(\alpha)$. It is very similar to the function R defined earlier except that the restraint is required to drop to 0 at stage σ if $\lim_{\tau \to \sigma} g'(\tau,e) \neq g'(\sigma,e)$. We now set $R(\sigma,\beta) = 0$ if either this happens or if $L(\sigma,g'(\sigma,\beta)) = M(\sigma,g'(\sigma,\beta))$. Let $R(\sigma,\beta)$ be

$$\sup \{\tau \mid \tau < \sigma \ \& \ L(\tau,g'(\sigma,\beta)) = M(\tau,g'(\sigma,\beta))\} + 1$$

otherwise. Let $R^*(\sigma,\beta) = \sup \{R(\sigma,\beta') \mid \beta' < \beta\}$.

We are now ready to describe the construction of the sets A_0 and A_1. We first set $A_i^{<\sigma} = \bigcup A_i^{\tau}$, $\tau < \sigma$. At stage σ find the least β and the least i such that if $g'(\sigma,\beta) = e$ then

(7.9) $W_e^{\sigma} \cap A_i = 0$.

(7.10) There is an x in $W_e^{\sigma} > R^*(\sigma,\beta)$ such that if $\{g'(\sigma,\beta)\}_{\tau}^{A_i^{\tau}} | \gamma$ is injured by putting x in A_i, then this computation has not been injured thus far by a requirement $R_{e'}(A_j)$ whose priority at stage σ is higher than that of $R_e(A_i)$.

(7.11) The ordinal x in (7.10) is greater than $\sup \{\tau | \tau < \sigma \ \& \ g'(\tau,\beta) \neq \lim_{\nu \to \tau} g'(\nu,\beta)\}$.

(7.12) $\lim_{\tau \to \sigma} g'(\tau,\beta) = g'(\sigma,\beta)$.

Put the least x satisfying (7.10) and (7.11) in A_i, where β, e are the least ordinals satisfying the above.

<u>7.13 LEMMA</u> Let K be an α-finite subset of $\sigma 2p(\alpha)$. Suppose that for each β' in K, $K(\beta') = \{\sigma | R(\sigma,\beta') = 0\}$ is unbounded. Then $K(\beta')$ is also (set-theoretically) closed, so that $\cap K(\beta')$, $\beta' \in K$, is closed and unbounded (it is clearly α-recursive).

Proof. Suppose $R(\sigma,\beta') = 0$ for σ cofinal in a limit ordinal λ. There are two possibilities. The first is when $\lim_{\sigma \to \lambda} g'(\sigma,\beta') \neq g'(\lambda,\beta')$. Then we automatically have $R(\lambda,\beta') = 0$. The other possibility is when there is a $\sigma^* < \lambda$ such that $g'(\sigma,\beta') = g'(\lambda,\beta')$ for $\sigma^* < \sigma < \lambda$. For each such σ, if $R(\sigma,\beta') = 0$, then we have $L(\sigma,g'(\sigma,\beta')) = M(\sigma,g'(\sigma,\beta'))$. Between σ and λ, only finitely many stages τ take place where $L(\tau,g'(\tau,\beta')) < M(\sigma,g'(\sigma,\beta'))$, after which the inequality $<$ becomes $\geq$. We must then have $L(\lambda,g'(\lambda,\beta')) = M(\lambda,g'(\lambda,\beta'))$. Thus $K(\beta')$ is closed and unbounded. The second conclusion now follows.

7.14 LEMMA If W_e is unbounded in α, then $W_e \cap A_i \neq 0$.

Proof. Let β be the least δ such that $g(\delta) = e$. Proceed to a stage σ_0 where $g'(\sigma,\beta') = g'(\sigma_0,\beta')$ for all $\beta' < \beta$ in the domain of g and for all $\sigma > \sigma_0$ (such σ_0 exists by Σ_2 admissibility). Let K be the set $\{\beta' | \beta' < \beta \ \& \ \beta' \in \mathrm{dom}(g)\}$. K is α-finite. Assume that σ_0 has also been chosen so that if β' is in K and there is an x in $W_{g(\beta')} \cap A_j$ then it already occurs by stage σ_0. Choose $\sigma_1 > \sigma_0$ such that every computation $\{g(\beta')\}_\tau^{A_i^\tau} | \gamma$, $\beta' \in K$, which gets injured is reinstated by stage σ_1, if it ever does. As in the case for $\sigma 2p(\alpha) = \alpha$, none of these gets injured after stage σ_1. Let $K_0 = \{\beta' | \beta' < \beta \ \& \ R(\sigma,g'(\sigma,\beta')) = 0$ for unboundedly many $\sigma\}$. Let $K_1 = \beta - K_0$. Both K_0 and K_1 are α-finite. Then $\{\sigma | L(\sigma,g'(\sigma,\beta')) = M(\sigma,g'(\sigma,\beta'))$ for some β' in $K_1\}$ is bounded by a $\sigma_2 > \sigma_1$ (observe that K_1 is bounded in $\sigma 2p(\alpha) < \sigma 2cf(\alpha)$). By Lemma 7.13, the set of stages $\sigma > \sigma_2$ where $R(\sigma,g'(\sigma,\beta')) = 0$ for β' in K_0 is α-recursive, closed and unbounded. Moreover, for each δ in $\beta - K$, the set of stages σ such that $\lim_{\tau \to \sigma} g'(\tau,\delta) \neq g'(\sigma,\delta)$ is closed and unbounded. Let X be the intersection of these closed and unbounded sets. It is again α-recursive and closed and unbounded. Let $x > \sigma_2$ be an element of W_e. Then x would enter A_i at a stage $\sigma > \sigma_2$ in X ((7.12) is satisfied).

7.15 LEMMA If $\{e\}^{A_0} = \{e\}^{A_1} = C$, then C is α-recursive.

Proof. Let $g(\beta) = e$. Let σ_0 be chosen so that for all $\beta' < \beta$ and β' in dom(g), $g'(\sigma,\beta') = g'(\sigma_0,\beta')$ for each $\sigma > \sigma_0$. Let K_0 be the set containing all such β'. Then K_0 is α-finite. Choose $\sigma_1 > \sigma_0$ such that for β' in K_0, if $W_{g(\beta')} \cap A_j \neq 0$, then it already occurs by stage σ_1. As in the proof of Lemma 7.8, for each γ there is a stage $\sigma(\gamma) > \sigma_1$ where $L(\sigma(\gamma),g'(\sigma(\gamma),\beta)) = M(\sigma(\gamma),g'(\sigma(\gamma),\beta)) > \gamma$. This is proved by invoking the regularity of A_i (a consequence of $\sigma 2cf(\alpha) = \alpha$). By Lemma 7.13 the set X of σ's such that $L(\sigma,g'(\sigma,\beta)) = M(\sigma,g'(\sigma,\beta))$ is α-

recursive, closed and unbounded. Also for each β' in $\beta - K_0$, the set $X(\beta')$ of σ's such that $\lim_{\tau \to \sigma} g'(\tau,\beta') \neq g'(\sigma,\beta')$ is α-recursive, closed and unbounded. Then $Y = X \cap (\cap X(\beta'))$, for β' in $\beta - K_0$, is again α-recursive, closed and unbounded. Now fix γ. To compute $C|\gamma$, choose a stage σ_2 in Y greater than σ_1 such that $L(\sigma_2, g'(\sigma_2,\beta))$ is equal to $M(\sigma_2, g'(\sigma_2,\beta))$. By the choice of σ_2, no requirement $R_{e'}(A_j)$ where $g'(\sigma,\beta') = e'$, β' in $\beta - K_0$ and $\sigma > \sigma_1$, will ever injure a computation of the form $\{e\}^{A_0^{\sigma 2}}_{\sigma 2}|\gamma$, $\{e\}^{A_1^{\sigma 2}}_{\sigma 2}|\gamma$. On the other hand, at most one of the two computations is injured at stage $\sigma > \sigma_2$ by a requirement $R_{e'}(A_j)$ of lower priority than S_e (at stage σ). One now argues as in Lemma 7.8 that $C|\gamma$ can be effectively computed. This completes the proof of the lemma as well as the proof of Theorem 7.1.

<u>7.16 REMARK</u> Lerman and Sacks [1972] showed that minimal pairs exist if α is not refractory, i.e. if it is not the case that $\sigma 2p(\alpha)$ = greatest α-cardinal $< t\sigma 2p(\alpha)$. The result proved above (Shore [1978]) overlaps with the Lerman-Sacks result, and solves a case left open in their paper. The only unsolved case currently for the minimal pair problem is when $\sigma 2cf(\alpha) < \sigma 2p(\alpha)$ = greatest α-cardinal $< \alpha$.

CHAPTER 8

THE DENSITY THEOREM

The technique used in the construction of minimal pairs discussed in the last chapter is an example of (a weak form of) infinite injury priority argument in α-recursion theory. This chapter studies the technique deeper by proving the Density Theorem: If $\mathbf{d} < \mathbf{c}$ are two degrees of α-r.e. sets, then there is a degree $\mathbf{a}$ of an α-r.e. set such that $\mathbf{d} < \mathbf{a} < \mathbf{c}$. The celebrated result for $\alpha = \omega$ was proved by Sacks ([1964]). A complete generalization of this theorem to all α was regarded to be a significant test for the success of α-recursion theory as a form of generalized recursion theory. The first density result in higher recursion theory was obtained by Driscoll ([1968]), when he showed that the r.e. degrees in ω_1^{CK} are dense. Shore [1976], exploiting his method of blocking (cf. Chapter 3), obtained a solution for all admissible ordinals α. The proof to be given below will be different from that of Shore [1976], benefiting from Soare's beautiful exposition [1976] of the infinite injury method (now more appropriately called <u>the $\mathbf{0}''$-infinite injury priority method</u>).

We begin with an outline of the proof. Let $D <_\alpha C$ be regular α-r.e. sets. By Theorem 1.22 regular α-r.e. sets exist in every α-r.e. degree. The α-r.e. set A to be constructed will consist of triples (e,x,τ). Let $A^{(0)}$ be the set of all (x,τ)'s such that $(0,x,\tau) \in A$. Then $A^{(0)} = D$. In this manner we code D into A and so automatically D is α-recursive in A. This is of course easy to do. The next step is to introduce two types of requirements, known as positive (P_e) and negative (N_e) requirements:

(P_e): $\{e\}^D \neq A$, and

(N_e): $\{e\}^A \neq C$.

We consider for our purposes only relative α-recursiveness. In other words, the expressions $\{e\}^D \neq A$ and $\{e\}^A \neq C$ assert respectively that A is not α-recursive in D via the reduction procedure e and C is not α-recursive in A via the reduction procedure e. The possibility that, for example, A may be weakly α-recursive in D via the reduction procedure e is not excluded. By considering only the former notion of reduciblity, the construction and proof that A is of intermediate degree is considerably simplified.

Now satisfying these two requirements constitutes the reason for the intricate technical devices introduced in the construction. As is typical of α-recursion theory, new problems arise when an admissible ordinal is to perform a job that seems to require more than Σ_1 admissibility. In this case the problem is much more acute since $\Sigma_3(L_\alpha)$ functions are involved, so that the various $\Sigma_2(L_\alpha)$ parameters introduced in the previous chapters, such as $\sigma 2p(\alpha)$ and $\sigma 2cf(\alpha)$, are no longer sufficient to handle the complexities arising from the construction. We elaborate on this by first recalling the proof of the classical case. Familiarity with the proof as presented in Soare [1976] will help towards understanding the discussion below.

Fix α to be an infinite regular cardinal, unless otherwise specified. The elements of A come in triples (e,x,τ). We introduce the following notations:

$$(8.1)\qquad a^\sigma = \begin{cases} \text{the least element } x \text{ in } A^\sigma - A^{<\sigma}, & \text{if } x \text{ exists} \\ \sup\,(A^{<\sigma} \cup \{\sigma\}) & \text{otherwise.} \end{cases}$$

(8.2) (Use function)

$$U(\sigma,e,x) = \begin{cases} \text{the least } z \text{ such that } \{e\}_\sigma^{A^{<\sigma}|z}(x) \text{ is defined} & \text{if } z \text{ exists} \\ 0 & \text{otherwise.} \end{cases}$$

Thus the use function determines how much information about A is used in the computation of $\{e\}^A(x)$ at stage σ. As usual the idea is to protect $A^{<\sigma}|z$ as much as possible, so that computations obtained using A as an oracle will not be destroyed.

If σ is a limit ordinal, let $u(\sigma)$ be the least u for which there is a sequence $\{x_\zeta\}$, $\zeta < \sigma$, of ordinals cofinal in u such that $x_\zeta \in A^\zeta - A^{<\zeta}$. Let $u^*(\sigma) = \min\{u(\sigma), a^\sigma\}$. Let $A(\sigma) = A^{<\sigma}|u^*(\sigma)$.

(8.3) $$\{e^\wedge\}_\sigma^{A^{<\sigma}}(x) = \begin{cases} \{e\}_\sigma^{A^{<\sigma}}(x) & \text{if this is defined } \sigma = \sigma'+1, \text{ and } U(\sigma,e,x) < a^\sigma. \\ \{e\}_\sigma^{A(\sigma)}(x) & \text{if } \sigma \text{ is a limit ordinal, this is defined, and } U(\sigma,e,x) < u(\sigma). \\ \text{undefined} & \text{otherwise.} \end{cases}$$

(8.4) (Modified use function)

$$\hat{U}(\sigma,e,x) = \begin{cases} U(\sigma,e,x) & \text{if } \{e^\wedge\}_\sigma^{A^{<\sigma}}(x) \text{ is defined.} \\ 0 & \text{otherwise.} \end{cases}$$

(8.5) (Length of agreement function)

$$L^\wedge(\sigma,e) = \text{the least } x \text{ such that } C^\sigma(x) \neq \{e^\wedge\}_\sigma^{A^{<\sigma}}(x).$$

The length of agreement function differs slightly from that given in the previous chapter, in that the reduction procedure $e^\wedge$ is used instead of e. The introduction of the $^\wedge$ function, introduced by Lachlan, simplifies the construction in a rather surprising way.

(8.6) (Modified length of agreement function)

Let $s = \sup\{x \mid (\exists \tau<\sigma)(x < \hat{L}(\tau,e)$ &

$(\forall y<x)(A^{<\sigma}|U(\sigma,e,y) = A^{<\tau}|U(\tau,e,y))\}$.

Then $\hat{M}(\sigma,e) = s$ if s is a limit ordinal, and equal to s+1 otherwise.

Thus the modified length of agreement function identifies the maximum length where information about A used for the computation of $\{e\}^A(y)$ at a stage $\tau < \sigma$ remains unchanged at stage σ, regardless of whether $C^{<\sigma}(y) = C^{<\tau}(y)$.

(8.7) (Restraint function)

$\hat{R}(\sigma,e) =$ the least $z > \hat{U}(\sigma,e,x)$, for $x < \hat{M}(\sigma,e)$.

Warning $\hat{L}(\sigma,e)$, $\hat{M}(\sigma,e)$, $\hat{R}(\sigma,e)$ defined in (8.5) - (8.7) will be given different definitions later on when we consider an arbitrary admissible ordinal α. They coincide with the present ones wher α is an infinite regular cardinal.

As usual no element will be allowed to enter A to satisfy a requirement (positive or negative) unless it lies above the restraints imposed by all requirements of higher priority.

(8.8) (Injury set)

$\hat{I}(\sigma,e) = \{x \mid (\exists\tau<\sigma)(x < \hat{R}(\sigma,e)$ & $x \in A^{\sigma} - A^{<\sigma}\}$.

We let $\hat{I}(e) = \bigcup \hat{I}(\sigma,e)$, $\sigma < \alpha$.

Finally, define $\hat{L}_D$ as for $\hat{L}$, except that we replace A by D and C by A in (8.5).

A triple (e,x,τ) is enumerated into A at stage σ if $x \in C^{\sigma}$,

(8.9) $(e,x,\tau) > \hat{R}(\sigma,d)$ for all $d < e$, and

(8.10) $x < \hat{L_D}(\rho,e)$ for all ρ such that $\tau \leq \rho \leq \sigma$.

To argue that the set A constructed has intermediate degree, one shows by induction that

(8.11) $A^{(<e)} = \{\langle x,\tau\rangle \mid (\exists d)(0 < d < e \ \& \ (d,x,\tau) \in A\}$ is recursive. It will then follow that $\bigcup \hat{I}(d)$, $d < e$, is recursive as well.

(8.12) $\liminf_\sigma \hat{L}(\sigma,e) < \alpha$; hence C is not α-recursive in A via the reduction procedure e.

(8.13) Let $T = \{\sigma \mid A^{<\sigma} | a^\sigma = A | a^\sigma\}$, the 'truth set' of A (these are the stages where correct guesses about $A|a^\sigma$ is made; observe that T is Turing equivalent to A). If $\liminf_\sigma \hat{L}(\sigma,e) < \alpha$, then we have $\lim_{\sigma \in T} \hat{R}(\sigma,e) < \alpha$.

(8.14) $\liminf_\sigma \hat{L_D}(\sigma,e) < \alpha$; hence A is not recursive in D via the reduction procedure e.

(8.15) $A^{(e)} = \{(x,\tau) \mid (e,x,\tau) \in A\}$ can be effectively computed from C via some $\{i(e)\}^C$, where i(e) can be recursively computed from C.

(8.12) and (8.14) together ensure that the positive and negative requirements are satisfied, while (8.15) implies that A is recursive in C. (8.11) and (8.13) are intermediate steps.

In generalizing the Density Theorem to arbitrary admissible ordinals α, we first fix $D <_\alpha C$ to be regular α-r.e. sets. In the construction of A, we replace individual requirements by blocks of requirements, as was done for the Splitting Theorem (Theorem 3.26). Each block of requirements will be accorded the same priority. While the number of blocks used in the proof of Theorem 3.26 was equal to the Σ_2

cofinality of α, this ordinal is not suitable for our purposes here. Instead we take the $\Sigma_2(<L_\alpha,D>)$ cofinality to be the number of blocks of requirements. This will be used to satisfy analogs of (8.12) and (8.14). It will also turn out that the values of lim $\inf_\sigma \hat{L}(\sigma,\gamma)$, lim $\inf_\sigma \hat{R}(\sigma,\gamma)$ and lim $\inf_\sigma \hat{L_D}(\sigma,\gamma)$ (for the γth block of requirements) are Σ_2 in $<L_\alpha,D>$ and bounded in α. This allows one to argue that C is not α-recursive in A and A is not α-recursive in D via any reduction procedure in the γth block. The choice of the ordinal $\Sigma_2(<L_\alpha,D>)$ is designed to ensure that each block is eventually satisfied in this sense.

Now it is also technically necessary to give the requirements a short indexing list. Here we take the Σ_1 projectum of $<L_\alpha,D>$ to form the list. This allows one to conclude that every $\Sigma_1(<L_\alpha,D>)$ set bounded below $\sigma 1p(\alpha,D)$ is α-D-finite (i.e. an element of $<L_\alpha,D>$), and hence α-finite (since D is a regular α-r.e. set)). One more ordinal will be needed in the course of the construction. It is the Σ_1 cofinality of $<L_\alpha,D>$ (also known as the recursive cofinality of D). The idea of introducing these parameters for the Density Theorem is due to Shore [1976]. For the rest of this chapter, we use the following notations:

$$\alpha^*(D) = \sigma 1p(\alpha,D)$$

$$\beta_1 = \sigma 1cf(\alpha,D)$$

$$\beta_2 = \sigma 2cf(\alpha,D)$$

Let $f: \alpha^*(D) \to \alpha$ be a partial α-D-recursive function (i.e. a partial function which is $\Sigma_1(<L_\alpha,D>)$) from a subset of $\alpha^*(D)$ onto α (cf. Chapter 1 for the discussion of the relativizations of projecta). Let g_1 be a $\Sigma_1<L_\alpha,D>)$ cofinality function from β_1 into α. Let g_2 be a $\Sigma_2(<L_\alpha,D>)$ cofinality function from β_2 into α. We will later define a particularly nice $\Sigma_2(<L_\alpha,D>)$ cofinality function $g_2{}^*$ for our construction. Our strategy is to divide $\alpha^*(D)$ into β_2 many blocks, using a $\Sigma_2(<L_\alpha,D>)$ cofinal function $H: \beta_2 \to \alpha^*(D)$ obtained by combining f with $g_2{}^*$. The verification that A has intermediate α-degree now proceeds by induction on blocks of requirements . One shows first of all that for each $f^{-1}(e) < \alpha^*(D)$, the set

$$A^{(<e)} = \{(x,\tau) \mid (\exists d)(f^{-1}(e) > f^{-1}(d) > 0 \; \& \; (d,x,\tau) \in A\}$$

is α-recursive, from which one argues that for each γ, C is not recoverable from A even with the help of the set of all reduction procedures in the γth block. By appealing to the coding strategy embodied in the construction, one shows that neither is A recoverable from D using the set of all reduction procedures in the γth block. Both of these rely on the fact that $D <_\alpha C$.

We begin the proof of the Density Theorem by first defining some paticularly nice α-recursive approximations for the functions f, g_2 and g_1.

Let c_f be chosen so that $f(x) = \{c_f\}^D(x)$ for all x in the domain of f, and $\{c_f\}^D(x)$ is undefined otherwise. An approximation f' of f is an α-recursive function such that

$$f'(\sigma,x) = \{c_f\}_\sigma^{D^{<\sigma}}(x).$$

As D is regular, we see that if f(x) is defined, there is a σ_0 such that for all $\sigma > \sigma_0$,

$$(8.16) \qquad f'(\sigma,x) = \{c_f\}_\sigma^{D^{<\sigma}}(x).$$

Next we let $g_1 = \{c_{g_1}\}^D$, so that again there is an α-recursive function $g_1{}'$ such that $\lim_\sigma g_1{}'(\sigma,x) = \{e_{g_1}\}^D(x)$ for all $x < \beta_1$. Now it is possible to choose g_1 and $g_1{}'$ such that g_1 is strictly increasing and $g_1'(\sigma,x) \geq g_1'(\tau,y)$ whenever $\sigma \geq \tau$ and $x \geq y$. It is also possible to choose g_1 to be continuous, i.e. $g_1(x) = \lim g_1(y)$ if $y \to x$. Similarly one may also choose $g_1'(\sigma,x) = \lim g_1'(\sigma,y)$ whenever $y \to x$. We assume that all of these are done. Now by the fact that g_1 is defined on β_1 we have

(8.17) For all $x < \beta_1$, there is a $\sigma(x)$ such that

$$g_1'(\sigma,y) = g_1'(\sigma(x),y) \text{ for all } \sigma \geq \sigma(x) \text{ and } y \leq x,$$

i.e. the function g_1 is 'tame'.

We also have an approximation g_2' for g_2. In this case $g_2'(\sigma,x) = \{e_{g_2}\}^D(\sigma,x)$, and $g_2(x) = \lim_\sigma g_2'(\sigma,x) = \lim_\sigma \{e_{g_2}\}^D(\sigma,x)$ is defined for all $x < \beta_2$.

It is clear that $\beta_2 \leq \beta_1$. The following lemma shows that $\alpha^*(D) \leq \beta_1$.

<u>8.18 LEMMA</u> If $\beta_1 < \alpha^*(D)$, then D is a complete α-r.e. set. Hence (since $D <_\alpha C$) $\alpha^*(D) \leq \beta_1$.

Proof. Let B be a regular complete α-r.e. set. Suppose that B is defined by the $\Sigma_1(L_\alpha)$ formula ϕ, i.e. $x \in B$ if and only if $L_\alpha \models \phi(x)$. Let

$$K = \{(\gamma,\delta) \mid (\forall x<g_1(\gamma))(x \in B \leftrightarrow x \in B^{g_1(\delta)}\}.$$

Unraveling the formula for the definition of K, we have

$$(\gamma,\delta) \in K \quad \leftrightarrow \quad (\forall w)(\forall z)(w = g_1(\gamma) \;\&\; z = g_1(\delta) \to$$
$$(\forall x<w)(x \in B \leftrightarrow x \in B^z).$$

Now K is a subset of $\beta_1 \times \beta_1$ which, being an α-cardinal, is closed under cartesian product and so bounded below

$\alpha^*(D)$. K is furthermore $\Pi_1(\langle L_\alpha, D\rangle)$, and so α-finite since $\beta_1 - K$ is $\Sigma_1(\langle L_\alpha, D\rangle)$. Now by the regularity of B, for each $\gamma < \beta_1$, there is a $\delta < \beta_1$ such that $(\gamma,\delta) \in K$. Since g_1 is weakly α-recursive in D, we have an oracle for deciding the set $B|g_1(\gamma)$ for each $\gamma < \beta_1$. Hence $B \leq_\alpha D$ and so D is a complete α-r.e. set.

<u>8.19 LEMMA</u> There is a total $\Sigma_1(\langle L_\alpha, D\rangle)$ map j from β_1 onto α. Furthermore, there exists an α-recursive approximation j' of j which is tame $\Sigma_2(L_\alpha)$.

Proof. We may assume that $\beta_1 < \alpha$. Then β_1 is an α-cardinal, hence closed under the pairing function. Let ϕ be a Δ_0 formula such that

$$f(x) = y \quad \leftrightarrow \quad \langle L_\alpha, D\rangle \models (\exists z)\phi(x,y,z).$$

$$\text{Let } j(x,\eta) = y \quad \leftrightarrow \quad [(x < \alpha^*(D) \;\&\; (\exists z<g_1(\eta)) \; (\langle L_\alpha, D\rangle \models \phi(x,y,z)] \text{ or } y = 0 \text{ and } [\ldots] \text{ fails.}$$

By Lemma 8.18, $\alpha^*(D) \leq \beta_1$. Hence j is total on β_1 and $\Delta_1(\langle L_\alpha, D\rangle)$. Since f is surjective and g_1 is unbounded in α, we see that j is surjective as well.

The approximation j': $\alpha\times\beta_1 \to \alpha$ is obtained by letting

$$j'(\sigma,(x,\eta)) = y \quad \leftrightarrow$$
$$[x < \alpha^*(D) \;\&\; (\exists z<g_1'(\sigma,\eta)) \; (\langle L_\alpha, D^\sigma\rangle \models \phi(x,y,z)]$$
$$\text{or } y = 0 \;\&\; [\ldots] \text{ fails.}$$

We leave it to the reader to verify that j' is indeed tame $\Sigma_2(L_\alpha)$.

The next lemma should be regarded as indicative of how the 'Σ_3-ness' of an infinite injury argument is reflected in admissible recursion theory through the use of approximation functions. Notice that the function $g_2{}^{*'}$ mentioned below attains the correct value of $g_2{}^*(x)$, for each $x < \beta_2$, at unboundedly many stages σ only (instead of for all sufficiently large σ as is the case for Σ_2 functions). The convergence of $g_2{}^{*'}$ to $g_2{}^*$ should be interpreted to mean that $g_2{}^*(x)$ is the least value equal to $g_2{}^{*'}(\sigma,x)$ for unboundedly many σ in α.

8.20 LEMMA There is an α-recursive function $g_2{}^{*'}$ approximating to a $\Sigma_2(\langle L_\alpha,D\rangle)$ function $g_2{}^*$ which maps β_2 cofinally into α such that

(a) $(\forall\gamma<\beta_2)(\exists\tau)(\forall x<\gamma)(\forall\sigma\geq\tau)(g_2{}^{*'}(\sigma,x) \geq g_2{}^*(x))$;

(b) $(\forall x<\beta_2)(\forall\tau)(\exists\sigma\geq\tau)(g_2{}^{*'}(\sigma,x) = g_2{}^*(x))$.

Proof. Let j be as in Lemma 8.19. Define

$$k(x,y) = 0 \quad \leftrightarrow \quad (\forall(\nu,\eta)\leq y<\beta_1)(\exists\sigma'\geq j(\eta))[\{e_{g_2}\}^D(\sigma',x) \neq j(\nu)].$$

Let $k(x,y)$ be undefined otherwise. Observe that $k(x,y)$ is defined if and only if the correct value of $g_2(x)$ has not been reached within the 'square' $\{(j(\nu),j(\eta)) \mid (\nu,\eta) \leq y\}$. The first y where $k(x,y)$ is undefined is also the first 'square' in which $\sigma(x)$ and $g_2(x)$ are found. Let x be fixed. Since $y < \beta_1$, and the clause [...] in the definition of $k(x,y)$ is weakly α-recursive in D, we conclude by an interchange of quantifiers that there is a σ' where for all $(\nu,\eta) \leq y$, there is a $\sigma'' \leq \sigma'$ such that $\sigma'' > \sup j(\eta)$ (where $(\nu,\eta) < y$ for some ν), and $\{e_{g_2}\}^D(\sigma'',x) \neq j(\nu)$. It follows that the function k is weakly partial α-recursive in D. Furthermore, for $\gamma < \beta_2 \leq \beta_1$, there is a y such that $k(x,y')$ is not defined for all $x < \gamma$ and $y' \geq y$. One also sees that $\beta_1 = \{y \mid k(x,y) = 0 \text{ for some } x < \beta_2\}$.

Let $\{e'\}^D(x,y) = 0$ if $(\exists w\leq x)(k(w,y) = 0)$, and let it be undefined otherwise. We define $g_2{}^{*'}$ by: $g_2{}^{*'}(\sigma,x) = g_1{}'(\sigma,y_{x,\sigma})$, where

$y_{x,\sigma}$ is the least z such that

> $\{e'\}_{\sigma}^{D^{\sigma}}(x,z)$ is undefined, or if $\sigma = \tau+1$, then the information used about D to compute $\{e'\}^{D}(x,z)$ at τ is discovered to be incorrect at stage σ.

Thus $y_{x,\sigma}$ is an approximation at stage σ to the least z such that $k(w,z)$ is undefined for all $w \leq x$. We set $g_2^*(x) = g_1(y_x)$, where y_x is this least z. Now the set of y_x's, for $x < \beta_2$, is unbounded in β_1 (by the way the function k is defined), and since g_1 is strictly increasing and cofinal in α, we see that g_2^* maps β_2 cofinally into α. Furthermore, since k is weakly α-recursive in D, g_2^* is a function which is $\Sigma_2(\langle L_\alpha, D\rangle)$.

Claim 1. Fix $\gamma < \beta_2$. There is a σ such that for all $x \leq \gamma$, for all $\tau \geq \sigma$, $g_2^{*\prime}(\tau,x) \geq g_2^*(x)$.

We have $g_2^*(x) = g_1(y_x)$, where y_x is the least z such that $k(w,z)$ is undefined for each $w \leq z$. For $x < \gamma$, we then have a $(\nu,\eta) < y_x$ $(< \beta_1)$ such that for all $\sigma' \geq j(\eta)$, $\{e_{g_2}\}^{D}(\sigma',x) = j(\nu)$. This means that for all $z < y_x$, $k(x,z)$ is defined. As $y_x < \beta_1$, and D is regular, there is a $\sigma(x)$ such that $\{e'\}_{\sigma(x)}^{D^{\sigma(x)}}(x,z) = 0$ for all $z < y_x$. Now we have $x \to \sigma(x)$ to be $\Sigma_2(\langle L_\alpha, D\rangle)$ and defined on $\gamma < \beta_2$. It follows that there is a stage σ bounding $\sigma(x)$ for all $x < \gamma$. For this σ, we have $g_2^{*\prime}(\tau,x) \geq g_2^*(x)$ for all $\tau \geq \sigma$.

Claim 2. For all $x < \beta_2$, for all σ, there is a $\tau \geq \sigma$ such that $g_2^{*\prime}(\tau,x) = g_2^*(x)$.

Choose $\gamma > x$, and let σ be as in Claim 1. Fix a $\sigma' \geq \sigma$. Then by Claim 1, for all $\tau \geq \sigma$, $g_2^{*\prime}(\tau,x) \geq g_2^*(x)$. If $g_2^{*\prime}(\tau_1,x) > g_2^*(x)$, then it must be that $\{e'\}_{\tau_1}^{D^{\tau_1}}(x,y_x) = 0$. By the definition of y_x and the regularity of D, there is a least $\tau \geq \tau_1$ such that $\{e'\}_{\tau}^{D^{\tau}}(x,y_x)$ is undefined. At this stage τ, we have $g_2^{*\prime}(\tau,x) = g_2^*(x)$.

The function $f: \alpha^*(D) \to \alpha$ which is a $\Sigma_1(\langle L_\alpha, D\rangle)$ surjection may be taken to be one-one. This is done as follows: Define a function h by induction. To compute h(x), use D to compute σ steps (for $\sigma < \alpha$) to see whether f(x) is defined. If and when it is defined, check to see whether f(x) = f(y) for some y defined at an earlier stage. This process ends before stage α since $\beta_1 > \alpha^*(D)$. If the answer is no, let h(x) = f(x). Otherwise h(x) is undefined. This gives a one-one surjection from a subset of $\alpha^*(D)$ onto α. Furthermore h is weakly α-recursive in D. From now on we assume that the function f is one-one.

8.21 LEMMA Let $H = f^{-1} \circ g_2^*$. Then H maps β_2 cofinally into $\alpha^*(D)$. Furthermore, H is $\Sigma_2(\langle L_\alpha, D\rangle)$.

Proof. Suppose that $H''\beta_2 \subset \delta < \alpha^*(D)$. Now $f''\delta$ is α-finite (since $\mathrm{dom}(f) \cap \delta$ is α-finite and $\alpha^*(D) < \beta_1$), hence bounded below α. But $g_2^{*}{}''\beta_2$ is unbounded in α, and $\delta \supset (f^{-1} \circ g_2^*)''\beta_2$. We have a contradiction. That H is $\Sigma_2(\langle L_\alpha, D\rangle)$ follows immediately from the fact that g_2^* is $\Sigma_2(\langle L_\alpha, D\rangle)$.

We now proceed to describe the construction of the α-r.e. set A. Notice first of all that the function H in Lemma 8.21 subdivides $\alpha^*(D)$ into β_2 many blocks. H is approximated by the α-recursive function H' defined to be

$$H'(\sigma,\gamma) = f^{-1\prime}(\sigma, g_2^{*\prime}(\sigma,\gamma)),$$

where $f^{-1\prime}$ is an α-recursive approximation to f^{-1}, just like f' is to f. Given a y, we have $f^{-1}(\sigma,y) = f^{-1}(y)$ for all sufficiently large σ. By Lemma 8.20 (a) and (b) we have

(8.22) $(\forall \gamma<\beta_2)(\exists \tau)(\forall \nu<\gamma)(\forall \sigma>\tau)(H'(\sigma,\nu) > H(\nu))$

(8.23) $(\forall \gamma<\beta_2)(\forall \tau)(\exists \sigma>\tau)(H'(\sigma,\gamma) = H(\gamma))$.

Clearly the stage τ in (8.22) tends to α as γ tends to β_2. It will be convenient, as a technical device, to impose on H' the following condition:

(8.24) If λ is a limit ordinal, then $H'(\lambda,\gamma)$ is the least y where a cofinal sequence of stages $\sigma < \lambda$ takes the value $H'(\sigma,\gamma) = y$ if such a y exists; otherwise let $H'(\lambda,\gamma) = \lambda$.

8.25 REMARK If $\sigma 2cf(\alpha) = \omega$, then $\beta_2 = \omega$ and so the function H is a $\Sigma_2(L_\alpha)$ function. In this case the function H' has the stronger property that

(8.26) For all $\gamma < \beta_2$, there is a τ such that $H'(\sigma,\nu) = H(\nu)$ for all $\nu < \gamma$ and all $\sigma \geq \tau$.

This fact will be used whenever $\beta_2 = \sigma 2cf(\alpha) = \omega$.

A block is a set of ordinals of the form

$$[0,H(\gamma+1)) = \{x \mid 0 \leq x < H(\gamma+1)\}$$

(called the γth block). A reduction procedure e is in the γth block if $f^{-1}(e) \in [0,H(\gamma+1))$. The γth block at stage σ is the set $[0,H'(\sigma,\gamma+1))$ (as usual we may arrange H to be strictly increasing, and H' to be non-decreasing). The reduction procedure e is in the γth block at stage σ if $f^{-1}{}'(\sigma,e)$ lies in the γth block at stage σ. The function H' will be used to guess at the block to which e actually belongs. As can be seen from (8.23) the guesses are correct unboundedly often.

We begin with a list of functions appropriately modified from (8.5) - (8.8). Define $\hat{e}$, $\hat{U}$ as before. We redefine the functions $\hat{L}(\sigma,e)$ and $\hat{L_D}(\sigma,e)$. This is prompted by the need to control very carefully the length of agreement of the reduction procedure e, to forestall the possibility that when D is not hyperregular, restraints are erroneously placed on an α-infinite amount of computations (associated with the reduction procedure e) which are defined only on a proper initial segment of ordinals. The idea here is to place restraints only on those computations which yield outputs on the initial segment $g_1(w)$ for $w < \beta_1$. It will then follow that if $\{e\}^A$ (or $\{e\}^D$ if we are considering $L_D\hat{}(\sigma,e)$) is not a total function, then only α-finitely much computations are given restraints (i.e. are 'protected'). This will allow the functions ('water marks') $R\hat{}(\sigma,e)$ and $R_D\hat{}(\sigma,e)$ to drop, so that elements may enter A to satisfy other requirements.

(8.27) $L\hat{}(\sigma,e) = g_1'(\sigma,w_e)$ where w_e is the greatest $w < \beta_1$ such that $\{e\hat{}\}_\sigma^{A^{<\sigma}}|g_1'(\sigma,w) = C^{<\sigma}|g_1'(\sigma,w)$ (recall that g' is continuous at every σ, so that the term 'greatest' is well-defined).

Define $L_D\hat{}(\sigma,e)$ in a similar manner, replacing A and C in (8.27) respectively by D and A. Next define $M\hat{}(\sigma,e)$ and $R\hat{}(\sigma,e)$ as in (8.6) and (8.7), except that $L\hat{}(\sigma,e)$ is now defined according to (8.27). Fix a limit stage λ. For each $\gamma < \beta_2$ and $\tau < \lambda$, let

(8.28)(a) $L'\hat{}(\tau,\gamma) = \sup L\hat{}(\tau,e)$, where the supremum is taken over those e in block γ at stage τ.

(b) $L\hat{}(\lambda,\gamma)$ = the least y such that for cofinally many $\tau < \lambda$, $L'\hat{}(\tau,\gamma) = y$ if such a y exists; otherwise let $L\hat{}(\lambda,\gamma)$ be $L'\hat{}(\lambda,\gamma)$ (defined as in (8.28) (a) with τ replaced by λ).

(8.29)(a) $M'\hat{}(\tau,\gamma) = \sup M\hat{}(\tau,e)$, where the supremum is taken over those e in block γ at stage τ.

(b) $M\ (\lambda,\gamma)$ be as defined in (8.28)(b), except that

$L'^{\wedge}(\tau,\gamma)$ is replaced by $M'^{\wedge}(\tau,\gamma)$.

(8.30)(a) $R'^{\wedge}(\tau,\gamma) = \sup R^{\wedge}(\tau,e)$, for e in block γ at stage τ.

(b) $R^{\wedge}(\lambda,\gamma)$ be as defined in (8.28)(b), except that $L'^{\wedge}(\tau,\gamma)$ is replaced by $R'^{\wedge}(\tau,\gamma)$.

(8.31)(a) $L_D'^{\wedge}(\tau,\gamma) = \sup L_D^{\wedge}(\tau,e)$, for e in block γ at stage τ.

(b) $L_D^{\wedge}(\lambda,\gamma)$ be as defined in (8.28)(b), with $L'^{\wedge}(\tau,\gamma)$ replaced by $L_D'^{\wedge}(\tau,\gamma)$.

At successor stage σ, let $L^{\wedge}(\sigma,\gamma)$, $M^{\wedge}(\sigma,\gamma)$, $R^{\wedge}(\sigma,\gamma)$ and $L_D^{\wedge}(\sigma,\gamma)$ be defined respectively as $L'^{\wedge}(\sigma,\gamma)$, $M'^{\wedge}(\sigma,\gamma)$, $R'^{\wedge}(\sigma,\gamma)$ and $L_D'^{\wedge}(\sigma,\gamma)$. Finally, for any stage σ, let

$$(8.32) \qquad I^{\wedge}(\sigma,e) = \{x \mid (\exists\tau<\sigma)(x < R^{\wedge}(\sigma,e) \ \& \ x \in A^{\tau} - A^{<\tau}\}.$$

Observe that $R^{\wedge}(\sigma,\nu) \leqslant R^{\wedge}(\sigma,\gamma)$ whenever $\nu \leqslant \gamma$.

For each e and each stage σ, let $\gamma(\sigma,e)$ be the least γ such that e is in block γ at stage σ (if defined). The construction at stage σ is as follows:

A triple (e,x,τ) enters A at stage σ if x is an element of C^{σ}, $x > R^{\wedge}(\sigma,\nu)$ for $\nu \leqslant \gamma(\sigma,e)$, and $x < L_D^{\wedge}(\rho,\gamma(\sigma,e))$ for all ρ such that $\tau \leqslant \rho \leqslant \sigma$.

<u>8.33 LEMMA</u> If for all $\sigma \geqslant \tau$, $H'(\sigma,\gamma) \geqslant H(\gamma)$, and if $\lambda \geqslant \tau$ is a limit ordinal such that $H'(\sigma,\gamma) = H(\gamma)$ cofinally in λ, then $H'(\lambda,\gamma) = H(\gamma)$.

Proof. Immediate from (8.24).

<u>8.34 LEMMA</u> A is regular.

Proof. Let $y < \alpha$. By the regularity of C let $\sigma_0 > y$ be a stage such that $C|y = C^{\sigma_0}|y$. Clearly one can choose an e such that $L\hat{}(\sigma,e) > y$ for all $\sigma > \sigma_0$. Let $\sigma_1 > \sigma_0$ be such that $f'(\sigma,u) = f(u)$ for all $u < f^{-1}(e)$ in the domain of f. Then for $(e',x',\tau') < y$, and for any $\sigma \geq \sigma_1$, if $f^{-1}{}'(\sigma,e') \geq f^{-1}(e)$ then $(e',x',\tau') \in A$ if and only if $(e',x',\tau') \in A^{\sigma_1}$. On the other hand, let $X = \mathrm{dom}(f) \cap f^{-1}(e)$. X is α-finite. For u in X, let

$$X_u = \{\sigma|R\hat{}(\sigma,f'(\sigma,u)) < R\hat{}(\sigma',f'(\sigma,u))$$
$$\text{for all } \sigma' < \sigma\}.$$

Each X_u is finite and $\{X_u\}$ is uniformly α-r.e. Since X is bounded in a regular α-cardinal $\leq \alpha^*(D)$, by Sacks-Simpson Lemma 3.1, $\bigcup X_u$ is α-finite. In particular, $\{\sigma(u)|\sigma(u) = \max X_u\}$ is α-finite. Then for $(e',x',\tau') < y$ such that $f^{-1}(e') < f^{-1}(e)$, we have $(e',x',\tau') \in A$ if and only if it is in $A^{\sigma(u)}$ for some u in X. This proves that A is regular.

Recall the definition of the truth set T of A in (8.13). We now show that T exists and has the same α-degree as A.

<u>8.35 LEMMA</u> The set T is unbounded in α and $T \equiv_\alpha A$. Furthermore for each $\gamma < \beta_2$, there are unboundedly many ordinals σ in α such that $H'(\sigma,\gamma) = H(\gamma)$ and either $\sigma \in T$ or $T \cap \sigma$ is unbounded in σ.

Proof. Let $y < \alpha$. By Lemma 8.34 choose σ_0 such that $A|y = A^{\sigma_0}|y$. Let x be the least ordinal greater than y enumerated in A after stage σ_0. Suppose $x = a^\sigma$. Then $A^{<\sigma}|a^\sigma = A|a^\sigma$.

To decide on $A|y$ from T, choose a σ in T such that $a^\sigma > y$. Then $A^{<\sigma}|a^\sigma = A|a^\sigma$, so that $A^{<\sigma}|y = A|y$. On the other hand, the fact that $T \leq_\alpha A$ is immediate.

Let $\gamma < \beta_2$. Suppose that for all $\sigma \geq \sigma_0$ such that

$H'(\sigma,\gamma) = H(\gamma)$, $\sigma \notin T$ and $T \cap \sigma$ is bounded in σ. First let $\sigma 2cf(\alpha) > \omega$. By (8.23) and (8.24) the set of σ's such that $H'(\sigma,\gamma) = H(\gamma)$ is a closed and unbounded, α-recursive set. Fix $\sigma(0) < \alpha$ such that $\sigma(0) >$ some τ which satisfies (8.22). For each $n \geq 1$, let $\sigma(n)$ be the least σ in T such that $\sigma > \sigma(n-1)$ and there is a ρ such that $H'(\rho,\gamma) = H(\gamma)$ and $\sigma(n-1) < \rho < \sigma$. The map $n \to \sigma(n)$ is $\Sigma_2(<L_\alpha>)$, and since $\sigma 2cf(\alpha) > \omega$, the supremum σ' of $\sigma(n)$, for $n < \omega$, is bounded below α. Then $H'(\sigma',\gamma) = H(\gamma)$ and $T \cap \sigma'$ is unbounded in σ'.

Now let $\sigma 2cf(\alpha) = \omega$. Then by (8.26) there is a stage $\tau < \alpha$ where $H'(\sigma,\gamma) = H(\gamma)$ for all $\sigma \geq \tau$. Choose a $\sigma > \tau$ such that σ is in T. This gives a σ required by the lemma.

To verify that the set A is of intermediate α-degree, we assume by induction on γ that

(8.36) $A^{(<H(\gamma))} = \{(x,\tau) \mid (\exists e)(H(\gamma) > f^{-1}(e) > 0 \ \& \ (e,x,\tau) \in A)\}$ is α-recursive.

(8.37) For each $\nu < \gamma$, $\lim_\sigma \hat{L}(\sigma,\nu) = L(\nu) < \alpha$, where σ is taken over those stages when $H'(\sigma,\nu+1) = H(\nu+1)$ and either $\sigma \in T$ or $T \cap \sigma$ is unbounded in σ.

By (8.22) there is a stage τ where $H'(\sigma,\nu'+1) \geq H(\nu'+1)$ for all $\nu' \leq \nu$ and $\sigma \geq \tau$. It follows for $\sigma \geq \tau$, we have $\hat{L}(\sigma,\nu)$ to be greater than or equal to $\sup\{\hat{L}(\sigma,e) \mid e \text{ in block } \nu\}$. On the other hand, if $\sigma \geq \tau$ and $\sigma \in T$ or $T \cap \sigma$ is unbounded in σ with $H'(\sigma,\nu+1) = H(\nu+1)$, then $\hat{L}(\sigma,\nu) \leq L(\nu)$. So $L(\nu)$ is the least ordinal ρ such that $\hat{L}(\sigma,\nu) = \rho$ for unboundedly many σ. In other words $\liminf_\sigma \hat{L}(\sigma,\nu) = L(\nu)$.

(8.38) For each $\nu < \gamma$, $\lim_\sigma \hat{L_D}(\sigma,\nu) = L_D(\nu) < \alpha$, where σ is taken over those stages as described in (8.37). Again $L_D(\nu)$ is the least ordinal ρ where $L_D\hat{}(\sigma,\nu) = \rho$ for unbloundedly many σ, i.e. $\liminf_\sigma L_D\hat{}(\sigma,\nu) = L_D(\nu)$.

For each e, let $a^{\sigma,e}$ be an enumeration of $A^{(e)} = \{(x,\tau) \mid (e,x,\tau) \in A\}$. Let $T^{(e)}$ be the set of all stages σ such that $(A^{(e)})^{<\sigma}|a^{\sigma,e} = (A^{(e)})|a^{\sigma,e}$. Define $T^{(<H(\nu))}$ similarly, replacing $A^{(e)}$ everywhere by $A^{(<H(\nu))}$. We leave it to the reader to prove the appropriate analogs of Lemma 8.35.

<u>8.39 LEMMA</u> For each $\nu < \gamma$, $\alpha > R(\nu) = \lim_\sigma R\hat{}(\sigma,\nu)$, for σ such that $H'(\sigma,\nu+1) = H(\nu+1)$ and either σ in $T^{(<H(\nu))}$ or $T^{(<H(\nu))} \cap \sigma$ is unbounded in σ. Furthermore, $\nu \to R(\nu)$ is a $\Sigma_2(<L_\alpha,D>)$ map.

Proof. We do this by induction on $\nu < \gamma$. Suppose this is true for all $\nu' < \nu$. As $\nu < \beta_2$, $\{R(\nu') \mid \nu' < \nu\}$ is bounded below α by a (least) ordinal y. Fix $L(\nu)$ as given in (8.37) and let τ_0 be a τ satisfying (8.22) when the γ there is set to be $\nu+1$. Let $\tau_1 > \tau_0$ such that $C|L(\nu) = C|^{\tau_1}L(\nu)$ (the stage τ_1 exists by the regularity of C). We may suppose that $\tau_1 > y$. By the definition of $L\hat{}$ and the fact that g_1 is continuous, we have for each stage σ, $L\hat{}(\sigma,\nu) = g_1(x)$ for some $x < \beta_1$. In particular $L(\nu) = g_1(x_0)$ for some x_0. Without loss of generality, let us assume that for all $x < x_0$, and for all $\sigma > \tau_1$, $g_1'(\sigma,x) = g_1'(x)$.

<u>Claim.</u> There is a stage $\sigma^* > \tau_1$ where $H'(\sigma^*,\nu+1) = H(\nu+1)$, either σ^* is in $T^{(<H(\nu))}$ or $T^{(<H(\nu))} \cap \sigma^*$ is unbounded in σ^*, and where $L\hat{}(\sigma^*,\nu) = L(\nu)$.

By admissibility and (8.36), there are stages $\sigma > \tau_1$ where $H'(\sigma,\nu+1) = H(\nu+1)$, $L\hat{}(\sigma,\nu) = L(\nu)$, and such that the information used about $(A^{<\sigma})^{(<H(\gamma))}$ in the computation of $L\hat{}(\sigma,\nu)$ agrees with $A^{(<H(\nu))}$ (i.e. no wrong information about $A^{(<H(\nu))}$ is used). By the construction, if $L\hat{}(\sigma',\nu) \neq L(\nu)$ at some $\sigma' > \sigma$, it must be that wrong information about D (which forms $A^{(0)}$) was used at stage σ in the computation of $\hat{L}(\sigma,\nu)$). Now fix σ' such that either $\sigma' \in T$ or $T \cap \sigma'$ is unbounded in σ'. If the information used about D at stage σ' is correct, then we must have $\hat{L}(\sigma',\nu) < L(\nu)$. Since $A^{(<H(\nu))}$ is α-recursive, so is $T^{(<H(\nu))}$. For each $x < L(\nu)$, use D to find the least stage $\sigma(x)$ such that either $\sigma(x)$ lies in $T^{(<H(\nu))}$ or $T^{(<H(\nu))} \cap \sigma(x)$ is unbounded in $\sigma(x)$, with the following property: $\hat{L}(\sigma,\nu) \geq g_1(x)$, and the information used about D and $A^{(<H(\nu))}$ in the computation are correct. If σ^* as described in the claim does not exist, then $\sup \sigma(x)$, for $x < x_0$, is equal to α. Now the function $x \to \sigma(x)$ is weakly α-recursive in D, and so $\sigma lcf(\alpha,D) \leq x_0 < \beta_1$. This is a contradiction. Thus we may choose σ^* to be the least stage $\sigma \geq \sup \sigma(x)$ such that $H'(\sigma,\nu+1) = H(\nu+1)$, and either σ is in $T^{(<H(\nu))}$ or $T^{(<H(\nu))} \cap \sigma$ is unbounded in σ, to be the stage required by the claim.

By the choice of $L(\nu)$, we see that if $\hat{L}(\sigma,\nu) > L(\nu)$ for some $\sigma > \sigma^*$, then $\hat{L}(\sigma',\nu) = L(\nu)$ for some $\sigma' > \sigma$. Moreover we have $\hat{L}(\sigma,\nu) \geq L(\nu)$ and $\hat{R}(\sigma,\nu) \geq \hat{R}(\sigma^*,\nu)$ for all $\sigma \geq \sigma^*$.

As restraints are placed at stage σ only on computations (of reduction procedures e in block ν) on initial segments of the form $g_1(x)$, for $x < \beta_1$, we see that if $\hat{R}(\sigma',\nu) > \hat{R}(\sigma^*,\nu)$ for all $\sigma' > \sigma$ such that $H'(\sigma',\nu+1) = H(\nu+1)$, it must be that $\hat{L}(\sigma',\nu) > L(\nu)$ for all such σ. This of course contradicts the choice of $L(\nu)$. Hence there are unboundedly many stages $\sigma > \sigma^*$ where $H'(\sigma,\nu+1) = H(\nu+1)$ and $\hat{R}(\sigma,\nu) = \hat{R}(\sigma^*,\nu)$. This holds in particular whenever σ is either in $T^{(<H(\nu))}$ or $T^{(<H(\nu))} \cap \sigma$ is unbounded in σ. It follows that $R(\nu) = \hat{R}(\sigma^*,\nu) = \lim_\sigma \hat{R}(\sigma,\nu)$, where $H'(\sigma,\nu+1) = H(\nu)$ and either $\sigma \in T^{(<H(\nu))}$ or $T^{(<H(\nu))} \cap \sigma$ unbounded in σ, exists and is less than α.

Now to compute $R(\nu)$ consider only those stages σ where $H'(\sigma,\nu+1) = H(\nu+1)$. This can be done in a $\Sigma_2(\langle L_\alpha,D\rangle)$ manner since H is $\Sigma_2(\langle L_\alpha,D\rangle)$. Now suppose by induction that we have a function $r(\sigma,\nu')$ weakly α-recursive in D such that for some stage σ_0,

$r(\sigma,\nu') = R(\nu')$ for all $\sigma > \sigma_0$ and $\nu' < \nu$. Let h_1 be a function weakly α-recursive in D such that $\lim_{\sigma\to\alpha} h_1(\sigma,\mu) = H(\mu)$ for all $\mu < \beta_2$. Choose $\sigma_1 > \sigma_0$ such that $h_1(\sigma,\nu'+1) = H(\nu'+1)$ for all $\nu' < \nu$ and $\sigma > \sigma_1$. Now as was shown in the proof of the claim, one may use D and $A^{(<H(\nu))}$ to arrive at the stage σ^*. Then $R(\nu) = \hat{R}(\sigma^*,\nu)$. As $A^{(<H(\nu))}$ is α-recursive, the whole procedure is weakly α-recursive in D modulo the parameter $H(\nu+1)$, which is $\Sigma_2(\langle L_\alpha,D\rangle)$. Thus $\nu \to R(\nu)$ is Σ_2 in $\langle L_\alpha,D\rangle$. This completes the proof of Lemma 8.39.

We proceed with the inductive step.

8.40 LEMMA $L(\gamma) = \lim_\sigma \hat{L}(\sigma,\gamma)$, for σ such that $H'(\sigma,\gamma+1) = H(\gamma+1)$ and either $\sigma \in T^{(<H(\gamma))}$ or $T^{(<H(\gamma))} \cap \sigma$ is unbounded in σ, exists and is bounded below α. Hence for all e in block γ, C is not α-recursive in A via the reduction procedure e.

Proof. We first fix a τ as given in (8.22) for the ordinal $\gamma+1$. Let

$$\begin{aligned} X = \{f^{-1}(e) \mid (\exists z)(\exists\sigma>\tau)[&H(\gamma+1) = H'(\sigma,\gamma+1) \text{ \& e is in block } \gamma \text{ at stage } \sigma \text{ \& } \{\hat{e}\}_\sigma^{A^{<\sigma}}(z) = 0 \neq C^{<\sigma}(z) = 1 \text{ \&}\\ &(A^{<\sigma}|\hat{U}(\sigma,e,z))^{(0)} \text{ is an initial segment of D \&}\\ &(A^{<\sigma}|\hat{U}(\sigma,e,z))^{(<H(\gamma))} = (A|\hat{U}(\sigma,e,z))^{(<H(\gamma))}\}. \end{aligned}$$

Here '0' means 'not an element of' and '1' means 'an element of'. Let e be in X and let z and σ be chosen so that e belongs to X on account of the ordinals z and σ satisfying the definition of X. If $\sigma' > \sigma$, then $H'(\sigma',\gamma+1) \geq H'(\sigma,\gamma+1)$ by the choice of τ, and so e belongs to block ν for some $\nu \leq \gamma$. Now the computation $\{\hat{e}\}_\sigma^{A^{<\sigma}}(z)$ may be destroyed at stage σ' only if a triple (d,z',τ') enters A at stage σ', where $\gamma(\sigma',d) < \gamma$, $z' \in C^\sigma$, $\tau' \leq \sigma'$, and $(d,z',\tau') \leq \hat{R}(\sigma,\gamma)$. As we have $(A^{<\sigma}|\hat{U}(\sigma,e,z))^{(<\gamma)} = (A|\hat{U}(\sigma,e,z)^{(<\gamma)}$, such a σ' does not exist. It follows that indeed $\{\hat{e}\}^A(z) = 0 \neq C(z) = 1$. Thus all reduction procedures e in block γ which eventually verify $\{\hat{e}\}^A \neq C$ in this manner belong to X. Notice that X is $\Sigma_1(\langle L_\alpha, D\rangle)$, hence α-finite since it is bounded in $\alpha^*(D)$.

Suppose that $X = H(\gamma+1) \cap f^{-1}$. Define a function from X into α by sending each $f^{-1}(e)$ to the least z where together with a stage σ entitles one to put $f^{-1}(e)$ in X. This function is clearly seen to be weakly α-recursive in D, and so since $X = H(\gamma+1) \cap f^{-1}$ is bounded in $\alpha^*(D) < \beta_1$, the function is bounded in α by an ordinal y_0, say. Then there is a stage σ_0 such that $\sigma_0 \in T$ and for each e in block γ there is an $x_e < y_0$ such that $\{e\}^{A^{\sigma_0}}_{\sigma_0}(x_e) \neq C(x_e)$ and so the lemma will follow.

Now suppose that $H(\gamma+1) - X \neq \emptyset$. Let y_0 be an ordinal such that for all $f^{-1}(e)$ in X, the least z which for some stage σ puts $f^{-1}(e)$ in X lies below y_0. By the argument in the previous paragraph we know that y_0 exists. Assume that we have $\lim_\sigma \hat{L}(\sigma,\gamma) = \alpha$, where $H'(\sigma,\gamma+1) = H(\gamma+1)$ and either σ belongs to $T^{(<H(\gamma+1))}$ or $T^{(<H(\gamma+1))} \cap \sigma$ is unbounded in σ. We show that in this case $C \leq_\alpha D$ to obtain a contradiction. By (8.36), $I^*(H(\gamma)) = \bigcup I(e) = \bigcup I(\sigma,e)$, for $\sigma < \alpha$ and $f^{-1}(e) < H(\gamma)$, is contained in $A^{(<H(\gamma))}$ and so α-recursive as well. We now give an algorithm to compute C from D.

Let $x > y_0$ be given. To obtain $C|x$, use D to find a $\sigma > \tau$ such that $\hat{L}(\sigma,\gamma) > x$, $H'(\sigma,\gamma+1) = H(\gamma+1)$, and for e in block γ and $y_0 < x' < \hat{L}(\sigma,\gamma)$, if $y_0 < z \leq \hat{U}(\sigma,e,x')$ then either $z \notin I^*(H(\gamma))$ or $z \in A^\sigma$. Such a stage σ exists since $\lim_\sigma \hat{L}(\sigma,\gamma) = \alpha$, where $H'(\sigma,\gamma+1) = H(\gamma+1)$, and since we are concerned only with ordinals $z > y_0$ (we are also making implicit use of the fact that C and D are regular). Then for all $\sigma' > \sigma$ such that $H'(\sigma',\gamma+1) = H(\gamma+1)$, we have $\hat{L}(\sigma',\gamma) > \hat{L}(\sigma,\gamma)$ and

$$\hat{R}(\sigma',\gamma) > \sup\{\hat{U}(\sigma,e,x') \mid x' < x \ \&$$

$$e \in H(\gamma+1) - X\}.$$

This implies that for $e \in H(\gamma+1) - X$, and for x' such that $y_0 < x' < x$,

$$\{e^\wedge\}_\sigma^{A^\sigma}(x') = \{e\}^A(x') = C(x')$$

whenever the left hand side is defined (observe that for $\sigma > \tau$, an equality between $\{e\}_\sigma^{A^{<\sigma}}(z)$ and $C^{<\sigma}(z)$, for e in $H(\gamma+1) - X$ and $y_0 < z < x$, cannot turn into an inequality at a later stage since the 'A-side' is already fixed, while the 'C-side', if it changes value, can only change from '0' into '1', but then this would force e to be in X). This then provides us with an algorithm to compute $C(x)$ from D for every $x > y_0$. On the other hand $C|y_0$ is α-finite by the regularity of C. We have $C \leq_\alpha D$, a contradiction.

Now one obtains, following the argument in Lemma 8.39, the next result.

<u>8.41 LEMMA</u> $R(\gamma) = \lim_\sigma L^\wedge(\sigma,\gamma) < \alpha$, where σ is taken over those stages where $H'(\sigma,\gamma+1) = H(\gamma+1)$ and either σ belongs to $T^{(<H(\gamma))}$ or $T^{(<H(\gamma))} \cap \sigma$ is unbounded in σ.

<u>8.42 LEMMA</u> $L_D(\gamma) = \lim_\sigma L_D^\wedge(\sigma,\gamma)$, for σ such that $H'(\sigma,\gamma+1) = H(\gamma+1)$ and either $\sigma \in T^{(<H(\gamma+1))}$ or $T^{(<H(\gamma+1))} \cap \sigma$ is unbounded in σ, exists and is less than α. Hence for all e in block γ, A is not α-recursive in D via the reduction procedure e.

Proof. Again we select τ as given in (8.22) for $\gamma+1$ fixed. Let

$$Y = \{f^{-1}(e) \mid e \text{ in block } \gamma \ \& \ (\exists z)(\exists \sigma)[H'(\sigma,\gamma+1) = H(\gamma+1) \ \& \ \{e^\wedge\}_\sigma^{D^{<\sigma}}(z) = 0 \neq A^\sigma(z) = 1 \ \&$$

$$D^{\sigma}|U_D\hat{}(\sigma,e,z) = D|U_D\hat{}(\sigma,e,z)\}.$$

The set Y is contained in H(γ+1) hence bounded below $\alpha^*(D)$. We then have Y to be $\Sigma_1(<L_\alpha,D>)$ and bounded in $\alpha^*(D)$, hence α-finite. As in Lemma 8.40, if Y = H(γ+1) $\cap$ f^{-1} then Lemma 8.42 is proved. To see this, suppose that H(γ+1) $\cap$ f^{-1} = Y. Then by the fact that Y is bounded in $\alpha^*(D) < \beta_1$ we may use a function weakly α-recursive in D to pick a y_0 and a stage σ_0 such that for each $f^{-1}(e)$ in Y, there is a $z < y_0$ with $\{e\hat{}\}_{\sigma_0}^{D^{<\sigma_0}}(z) = 0 \neq A^{<\sigma_0}(z) = 1$. As A is α-r.e., an element once enumerated into A stays in it thereafter. The inequality displayed for each e in the definition of Y, using σ_0 and appropriate z, will then continue to hold at all $\sigma' > \sigma_0$. This then implies that $L_D(\gamma)$ as required by the lemma exists.

Thus let us assume that H(γ+1) $\cap$ f^{-1} - Y $\neq \emptyset$. Suppose that $L_D(\gamma)$ is unbounded in α. Then $\alpha = \lim_\sigma L_D\hat{}(\sigma,\gamma)$, where H'($\sigma$,$\gamma$+1) = H($\gamma$+1) and either $\sigma \in T^{(<H(\gamma))}$ or $T^{(<H(\gamma))} \cap \sigma$ is unbounded in σ. We provide an algorithm to compute C from D. By Lemma 8.41, there exist unboundedly many stages $\sigma > \sigma_0$ such that H'(γ+1) = H(γ+1) and $R\hat{}(\sigma,\gamma)$ = R(γ). Let $x > y_0$ and $x > R(\gamma)$. Let m(x) be the least stage $> \sigma_0$ such that $L_D\hat{}(\sigma',\gamma) > x$ for all $\sigma' > \sigma$ with H'(σ',γ+1) = H(γ+1). We claim that the ordinal m(x) exists.

For $f^{-1}(e)$ in H(γ+1) - Y, define $U_D\hat{}(\sigma,e,z)$ as for $U\hat{}(\sigma,e,z)$, except that A is replaced everywhere by D and we require that $\{e\hat{}\}_\sigma^{D^{<\sigma}}(z) = A^{<\sigma}(z)$. Then m(x) is the least σ such that H'(σ,γ+1) = H(γ+1), $L_D\hat{}(\sigma,\gamma) > x$, and $D^\sigma|u = D|u$, where u is the least ordinal greater than $U_D\hat{}(\sigma,e,z)$ for $y_0 < z < x$ and e an element of H(γ+1) - Y (as in the proof of Lemma 8.40 observe that for $\sigma > m(x)$, an equality between $\{e\hat{}\}_\sigma^{D^{<\sigma}}(z)$ and $A^{<\sigma}(z)$, for e in H(γ+1) - Y and $y_0 < z < x$, cannot turn into an inequality since the 'D-side' is already fixed, while the 'A-side', if it changes value, can only change from 'not an element of' into 'an element of', but then this would force e to be in Y). Now the map $x \to m(x)$ is weakly α-recursive in D, and we have, for each z such that $\max\{y_0, R(\gamma)\} < z < x$, $z \in C$ if and only if $(d,z,\rho) \in A$ for some d in block γ and some $\rho < m(x)$. To decide whether $(d,z,\rho) \in A$ one goes to the least stage $\sigma_1 > \sigma_0$ and greater than the supremum of the set $\{m((d,z,\rho)) \mid d,z,\ \rho$ as specified above$\}$. Then $u = (d,z,\rho)$ is in A if and only if for some e in $H(\gamma+1) - Y$, we have $\{e^{\wedge}\}^{D^{\sigma_1}}_{\sigma_1}(u) = A^{\sigma_1}(u) = 1$. One can clearly use this to obtain an algorithm to compute C α-recursively from D. As this is a contradiction, we have $L_D(\gamma)$ as specified in the lemma.

<u>8.43 REMARK</u> The reader should observe that Lemmas 8.40 and 8.42 do not claim that C is not weakly α-recursive in A, or that A is not weakly α-recursive in D, via the reduction procedure e (for all e in block γ). They only assert that no such e can provide an algorithm <u>for computing C α-recursively from A or A α-recursively from D.</u> The possibility that one may be <u>weakly α-recursive in</u> the other is not ruled out.

<u>8.44 LEMMA</u> For each $e > 0$ in block γ,

$$A^{(e)} = \{(x,\tau) \mid (e,x,\tau) \in A\}$$

is α-recursive.

Proof. By Lemma 8.42, $L_D(\gamma) < \alpha$. Then $\lim_\sigma L_D^{\wedge}(\sigma,\gamma) = L_D(\gamma)$ for σ such that $H'(\sigma,\gamma+1) = H(\gamma)$ and either $\sigma \in T^{(<H(\gamma))}$ or $T^{(<H(\gamma))} \cap \sigma$ is unbounded in σ. Let y be fixed. For x and τ such that $x \geq L_D(\gamma)$, with $(e,x,\tau) < y$, we have $(e,x,\tau) \in A$ only if it is in A^σ for the least stage $\sigma \geq \tau$ such that $H'(\sigma,\gamma+1) = H(\gamma+1)$ and $L_D^{\wedge}(\sigma,\gamma) = L_D(\gamma)$. For $x < L_D(\gamma)$, first fix σ' such that $C \mid L_D(\gamma) = C^{\sigma'} \mid L_D(\gamma)$. Then (e,x,τ) belongs to A only if it belongs to the least stage $\sigma'' > \sigma'$ such that $H'(\sigma'',\gamma+1) = H(\gamma+1)$, $\sigma'' > \tau$, and $R^{\wedge}(\sigma'',\gamma) = R(\gamma)$. This gives an algorithm for computing $A^{(e)}$ α-recursively.

8.45 LEMMA The set

$$A^{(<H(\gamma+1))} = \{(x,\tau) \mid (\exists e)(H(\gamma+1) > f^{-1}(e) > 0 \ \& \ (e,x,\tau) \in A\}$$
$$= \bigcup A^{(e)},\ 0 < f^{-1}(e) < H(\gamma+1)$$

is α-recursive.

Proof. By Lemma 8.44, each $A^{(e)}$, $e > 0$, is α-recursive. The proof of Lemma 8.44 actually uses only the parameters $f^{-1}(e)$, $L_D(\gamma)$ and $R(\gamma)$. Thus one may show $A^{(e)}$ to be α-recursive uniformly for all e where $0 < f^{-1}(e) < H(\gamma+1)$. Since $f^{-1} \cap H(\gamma+1)$ is α-finite, it follows that the set $A^{(<H(\gamma+1))}$ is α-recursive as well.

8.46 LEMMA There is a function $i: \alpha^*(C) \to \alpha$ weakly α-recursive in C such that $A^{(e)}$ is α-recursive in C via reduction procedure $i(f^{-1}(e))$.

Proof. Fix an ordinal ζ. Suppose that $i(f^{-1}(d))$ is defined for all $d < \zeta$ such that $f^{-1}(d) < f^{-1}(e)$. The set K of such d's is α-finite. By the fact that $f^{-1}(e) < \alpha^*(C) = \beta_1$, the set $\{i(f^{-1}(d))\}$ for d in K is α-finite and may be computed from C. It follows that $\bigcup T^{(d)}$, $d \in K$, is α-recursive in C. Let $y < \alpha$. By the regularity of C choose a stage (using C) σ_0 such that $C|y = C^{\sigma_0}|y$. Now suppose that (e,x,τ) is less than y. Of course if x is not in C then (e,x,τ) is not in A, and hence (x,τ) is not in $A^{(e)}$. Thus suppose that x is in C. Then x is in C^{σ_0}. Let $\sigma_1(x) > (e,x,\tau)$ be the least ζ greater than σ_0 such that for each d in K (as defined above), either $\zeta \in T^{(d)}$ or $T^{(d)} \cap \zeta$ is unbounded in ζ. The lemma would follow if we could show that the supremum s of the set $\{\sigma_1(x) \mid (e,x,\tau) < y \text{ for some } \tau\}$ is bounded in α. For then we have $(e,x,\tau) < y$ to be in A if and only if it is in the A by stage s. But the supremum exists according to Lemma 8.35, which implies that there is a stage $\sigma > \sigma_0$, y such that either $\sigma \in T$ or $T \cap \sigma$ is unbounded in σ. Clearly $\sigma > \sigma_1(x)$ for each $(e,x,\tau) < y$ (for some τ) and so $s \leq \sigma$. It follows that the ordinal $i(f^{-1}(e))$ is defined.

The following lemma takes care of the induction at limit γ.

<u>8.47 LEMMA</u> The maps $\gamma \to L(\gamma)$ and $L_D(\gamma)$ are $\Sigma_2(\langle L_\alpha,D\rangle)$.

Proof. This proceeds much the same as that given in Lemma 8.39. Namely, once the correct approximation to $H(\gamma+1)$ is reached (in a $\Sigma_2(\langle L_\alpha,D\rangle)$ manner), the computation of $L(\gamma)$ and $L_D(\gamma)$ from D then uses only $A^{(<H(\gamma))}$ (which is α-recursive) as parameters. The only thing to verify is that the parameters used to define $A^{(<H(\gamma))}$ can be obtained from D in a $\Sigma_2(\langle L_\alpha,D\rangle)$ manner. This however is a consequence of Lemma 8.45.

Finally, we show that A is α-recursive in C.

<u>8.48 LEMMA</u> $A \leq_\alpha C$.

Proof. Let $y_0 < \alpha$. Use f^{-1} to compute $f^{-1}(e)$ for each e such that $(e,x,\tau) < y_0$ for some x, τ. Use C to obtain a stage σ_0 where $C|y_0 = C^{\sigma_0}|y_0$ and $D|y_0 = D^{\sigma_0}|y_0$. By Lemma 8.45 and the construction, for $x < y_0$ in C, $(e,x,\tau) < y_0$ belongs to A if and only if it is already in A^{σ_1}, where σ_1 is the least stage $\zeta > \sigma_0$ such that if $d \leq \zeta$ and $f^{-1}(d) < f^{-1}(e)$, then either $\zeta \in T^{(d)}$ or $T^{(d)} \cap \zeta$ is unbounded in ζ. Such a σ_1 can be found from C (as in the proof of Lemma 8.46). Now to indicate the fact that σ_1 is a function of x and e, let us denote it as $\sigma_1(x)$. The lemma would follow if we could show that the supremum of $\{\sigma_1(x) | (e,x,\tau) < y_0 \text{ for some } \tau\}$ is bounded in α. But as in the proof of Lemma 8.46 this is a consequence of Lemma 8.35.

Using Lemmas 8.40, 8.42 and 8.48 we have

<u>8.49 THEOREM</u> The α-r.e. degrees are dense.

<u>8.50 REMARK</u> Together with the results in Shore [1976a], Maass [1978c] has used an infinite injury method to give a characterization of the existence of an incomplete high α-r.e. degree (i.e. an α-r.e. degree **a** such that $\mathbf{a}' = \mathbf{0}''$). This is divided into four cases:

(a) $\sigma 2cf(\alpha) \geq \sigma 2p(\alpha)$ and there is no incomplete non-hyperregular α-r.e. degree. These are precisely the ordinals which are Σ_2 admissible, and in this case there exists an incomplete high α-r.e. degree (Shore [1976a]).

(b) $\sigma 2cf(\alpha) \geq \sigma 2p(\alpha)$ and there is an incomplete non-hyperregular α-r.e. degree. Here we have $\alpha > \sigma 2cf(\alpha) \geq \sigma 2p(\alpha)$ and there exists an incomplete high α-r.e. degree.

(c) $\sigma 2cf(\alpha) < \sigma 2p(\alpha)$ and there exists an incomplete non-hyperregular α-r.e. degree. In this case there are only two α-degrees $\geq \mathbf{0}'$ (both less than $\mathbf{0}''$) which are the jumps of incomplete α-r.e. degrees. These are $\mathbf{0}'$ (the jump of all hyperregular α-r.e. degrees) and $\mathbf{0}^{3/2}$ (the jump of incomplete non-hyperregular α-r.e. degrees).

(d) $\sigma 2cf(\alpha) < \sigma 2p(\alpha)$ and there is no incomplete non-hyperregular α-r.e. degree. In this case $\mathbf{0}'$ is the jump of every incomplete α-r.e. degree (Shore [1976a]).

The degree $\mathbf{0}^{3/2}$ also exists in Case (b). It is the greatest $\Delta_2(L_\alpha)$ degree.

Compared to the α-finite injury method, the technique of infinite injury in the context of admissible recursion theory is less well understood.

CHAPTER 9

TREES

We have thus far devoted the study of admissible recursion theory almost exclusively to the analysis of the α-r.e. degrees and α-r.e. sets, introducing along the way various techniques that were invented to tackle problems arising therefrom. This chapter turns away from recursive enumerability and studies one of the most important problems in the structure theory of α-degrees: the existence of a minimal α-degree. An α-degree $\mathbf{a} > \mathbf{0}$ is minimal if for all α-degrees $\mathbf{b}$, if $\mathbf{b} < \mathbf{a}$, then $\mathbf{b} = \mathbf{0}$. The existence of a minimal ω-degree was first proved by Spector [1956]. By introducing priority method into Spector's construction, Sacks [1963b] constructed a set of minimal degree below $\mathbf{0}'$. McIntyre [1973] proved the existence of a minimal α-degree for all countable admissible ordinal α. Shore [1972] showed that minimal α-degrees exist for all Σ_2-admissible ordinals α, and Maass [1977] proved that this is true whenever $\sigma 2p(\alpha) \leq \sigma 2cf_\alpha(\alpha)$. We will prove the Maass-Shore result in this chapter.

The construction of a set of minimal α-degree hinges on the idea of a tree. To define this, we first introduce some basic notions.

9.1 DEFINITION A string σ is an α-finite function whose domain is an ordinal (less than α) and whose range is 2.

The length of a string σ, denoted $\mathrm{lth}(\sigma)$, is the least ordinal not in the domain of σ. We think of a string as the characteristic function of an α-finite initial segment of a subset of α (the convention is that for x in the domain of σ, $\sigma(x) = 1$ means x is in the set and $\sigma(x) = 0$ means x is not in the set). A string σ extends a string τ (written $\sigma \geqslant \tau$) if τ is the restriction of σ to $\mathrm{lth}(\tau)$. We use < to denote proper extension. Two strings are compatible if one extends the other, otherwise they are incompatible. The concatenation of two strings σ and τ is the string ν denoted $\sigma*\tau$ of length $\mathrm{lth}(\sigma)+\mathrm{lth}(\tau)$, such that for $x < \mathrm{lth}(\sigma)$, $\nu(x) = \sigma(x)$, and for $x = \mathrm{lth}(\sigma) + y$, $y < \mathrm{lth}(\tau)$, $\nu(x) = \tau(y)$. The notation $\sigma*i$, for $i = 0$ or 1, is used for $\sigma*\tau$ where $\mathrm{lth}(\tau) = 1$ and $\tau(0) = 0$ or 1 as the case may be. The symbol $\emptyset$ is used to denote the empty string.

<u>9.2 DEFINITION</u> A tree T is a partial α-recursive function from strings to strings such that

(i) T maps compatible strings to compatible strings.

(ii) If one of $T(\sigma*i)$, for $i = 0$ or 1, is defined, then so are all of $T(\sigma)$, $T(\sigma*0)$, $T(\sigma*1)$. In this case $T(\sigma*0)$ and $T(\sigma*1)$ are incompatible.

(iii) $\tau \leqslant \sigma$ and $T(\sigma)$ is defined implies that $T(\tau)$ is defined. In this case one also has $T(\tau) \leqslant T(\sigma)$.

(iv) If $\mathrm{lth}(\sigma)$ is a limit ordinal, and $T(\sigma)$ is defined, then it is the union of $T(\tau)$, $\tau < \sigma$. Conversely, if $T(\tau)$ is defined for all $\tau < \sigma$, and $\mathrm{lth}(\sigma)$ is a limit ordinal, then $T(\sigma) = \bigcup T(\tau)$, $\tau < \sigma$.

It follows from the above definition that T maps incompatible strings to incompatible strings, and that if σ properly extends τ then $T(\sigma)$ properly extends $T(\tau)$ (when $T(\sigma)$, hence also $T(\tau)$, is defined).

A string ρ lies on T if it is in the range of T. A set G is a branch of a tree T if for unboundedly many ordinals $\gamma < \alpha$, $G|\gamma$, the string of length γ which agrees with G on γ (we identify G with its characteristic function) lies on T. Clearly any set which is a branch of a tree is α-regular.

A tree U is a subtree of a tree T if every string that lies on U also lies on T.

Now let e be an index for a reduction procedure. We say that $T(\sigma)$ and $T(\tau)$ split $T(\nu)$ for e if $T(\sigma)$ and $T(\tau)$ extend $T(\nu)$ and there is an x such that $\{e\}^{\sigma}(x)$, $\{e\}^{\tau}(x)$ are both defined and are different. different. T is a splitting tree for e if for all strings σ, if $T(\sigma*0)$ and $T(\sigma*1)$ are defined then they split $T(\sigma)$ for e. The key idea in making G to be of minimal α-degree lies in ensuring that for every reduction procedure e, if $\{e\}^G$ is total, then there is a tree T containing G as a branch such that either T is a splitting tree for e, or there is a $\gamma < \alpha$ such that $G|\gamma$ lies on T and no pair of strings lying on T split $G|\gamma$ for e. This idea, of course, goes back to Spector [1956].

<u>9.3 LEMMA</u> Suppose that $\{e\}^G$ is the characteristic function of a set and G is a branch of a splitting tree T for e. Then G is α-recursive in $\{e\}^G$.

Proof. As noted above, since G is a branch of T, G is regular. We show that for all $\gamma < \alpha$, one may compute $G|\gamma$ effectively from $\{e\}^G$. Fix γ. Look for a string σ in the domain of T and an ordinal δ such that

(9.4) $\{e\}^{T(\sigma)}$ is compatible with $\{e\}^G$

(9.5) for all $\tau < \sigma$, if $\tau*i$ is incompatible with σ, then there is a $y < \delta$ such that at stage δ it is known that $\{e\}^{T(\sigma)}(y)$ is unequal to $\{e\}^{T(\tau*i)}(y)$ but is equal to $\{e\}^G(y)$.

(9.6) $\mathrm{lth}(T(\sigma)) \geqslant \gamma$.

Such a string σ exists by admissibility, the assumption that $\{e\}^G$ is a total function, and by the fact that T is a splitting tree. It then follows that $T(\sigma)$ is an initial segment of G. Since $lth(T(\sigma)) \geq \gamma$, we have computed $G|\gamma$ from $\{e\}^G$. This proves the lemma.

<u>9.7 LEMMA</u> Suppose that $\{e\}^G$ is the characteristic function of a set and G is a branch of a tree T. Suppose that there is an initial segment $G|\gamma$ that lies on the tree such that no pair of strings on T split $G|\gamma$ for e. Then $\{e\}^G$ is α-recursive.

Proof. Assume the hypothesis. To compute $\{e\}^G|\delta$, go to a string $T(\sigma)$ that lies on T such that $lth(\{e\}^{T(\sigma)})$ is at least δ and $T(\sigma) \geq G|\gamma$. This is an α-recursive search procedure, and so by admissibility and the assumption that $\{e\}^G$ is a characteristic function of a set (hence total), such a σ exists and can be α-recursively computed. Since no pair of strings on T extending $G|\gamma$ split $G|\gamma$ for e, we see that the computation $\{e\}^{T(\sigma)}|\delta$ actually agrees with $\{e\}^G|\delta$. Hence we conclude that $\{e\}^G$ is α-recursive.

Thus the problem of constructing a set of minimal α-degree is reduced to that of the construction of a set G which satisfies the conditions of either Lemma 9.3 or 9.7 for every reduction procedure e such that $\{e\}^G$ is a characteristic function ofa set. There is also the necessity of ensuring that G is not α-recursive. But as noted in Epstein and Posner [1978], the solution to the problem just described also entails the solution to this problem. We state this without proof.

<u>9.8 LEMMA</u> Let G be regular satisfying the following:

(9.9) For all reduction procedure e, if $\{e\}^G$ is the characteristic function of a set, then there is a tree T such that G is a branch of T where either T is a splitting tree for e, or there is an initial segment $G|\gamma$ of G lying on T for which no pair of strings lying on T and extending $G|\gamma$ split $G|\gamma$ for e. Then G is not α-recursive.

We construct a set G satisfying (9.9) by introducing a nested sequence of trees, and ensuring that the sequence has a nontrivial intersection, a branch of which will be the desired set G. The sequence is derived via two operations taking trees to trees.

9.10 DEFINITION Let τ be a string in the domain of a tree T. The splitting tree of T above $T(\tau)$ for the reduction procedure e is the tree $Sp(T,e,\tau)$ such that $Sp(T,e,\tau)(\emptyset) = T(\tau)$, and if $Sp(T,e,\tau)(\sigma) = T(\rho)$ has been defined, then $Sp(T,e,\tau)(\sigma*i)$ extends $T(\sigma*i)$, and $Sp(T,e.\tau)(\sigma*0)$, $Sp(T,e,\tau)(\sigma*1)$ split $T(\rho)$ for e. Furthermore this pair of strings, if exists, is the least pair (in terms of enumeration of elements of T) satisfying the above. Also if $lth(\sigma)$ is a limit ordinal, then $Sp(T,e,\tau)(\sigma)$ is the union of $Sp(T,e,\tau)(\rho)$, for $\rho \leq \sigma$.

9.11 DEFINITION Let τ be a string in the domain of a tree U. The full tree of U above $U(\tau)$ is the tree $Fu(U,\tau)$ such that $Fu(U,\tau)(\sigma) = U(\tau*\sigma)$.

$Fu(U,\tau)$ is introduced when no incompatible extensions of $U(\tau)$ split $U(\tau)$ for e. Specifically, suppose that we attempt to obtain $Sp(T,e,\rho)$ for some ρ in the domain of T. It may happen that Definition 9.10 is not satisfied, that is, there is a ν on $Sp(T,e,\rho)$ which has not proper extension on $Sp(T,e,\rho)$. This means that if $Sp(T,e,\rho)(\sigma) = \nu$, then there are no strings σ_0, σ_1 extending σ such that $Sp(T,e,\rho)(\sigma_i) > \nu*i$ and $Sp(T,e,\rho)(\sigma_i)$ split ν for e. It is of course possible that there may exist two strings $\nu'(j)$, $j = 0,1$, on $Sp(T,e,\rho)$ which extend $\nu*0$ (or $\nu*1$ as the case may be) which split ν for e. Thus suppose that we have for some x, $\{e\}^{\nu'(0)}(x) \neq \{e\}^{\nu'(1)}(x)$. This implies that for ν' on $Sp(T,e,\rho)$ extending $\nu*1$, $\{e\}^{\nu'}(x)$ is not defined (so that ifG extends $\nu*1$ and lies on $Sp(T,e,\rho)$, then $\{e\}^G$ is a partial function (hence not the characteristic function of a set). There is also the possibility that no splitting such as $\nu'(0)$, $\nu'(1)$ exists. In this case if we ensure that the desired set G extends $\nu*0$, then Lemma 9.7 will apply. In either case we will be able to conclude $\{e\}^G$ does not have α-degree strictly between 0 and that of G. The situation is entirely similar if we had chosen $\nu*1$ instead of $\nu*0$ in the above consideration. Thus in the definition of the full tree $Fu(U,\tau)$, we always choose U to be $Sp(T,e,\rho)$ for some T and τ to be such that either there is an ordinal x for which no extension of $U(\tau)$ on U converges on x (i.e. $\{e\}^{\nu'}(x)$ is undefined for all ν'on U extending $U(\tau)$ or no incompatible extensions of $U(\tau)$ on U split $U(\tau)$ for e.

A notion that is required in dealing with admissible ordinals $\alpha > \omega$ is that of recursive intersection of trees. This is provided by an α-recursive function denoted RI. Given a $\delta < \alpha$ and a sequence of trees $\{T_j\}_{j\varepsilon\delta}$ such that T_j is a subtree of $T_{j'}$ if $j' < j$ (we call this a nested sequence), we define $RI(\{T_j\})$ by induction. $RI(\{T_j\})(\emptyset) = \cup T_j(\emptyset)$, $j \varepsilon \delta$. If $RI(\{T_j\})(\sigma)$ is defined, and lies on T_j for every j (say $= T_j(\sigma_j)$), then $RI(\{T_j\})(\sigma*i) = \cup T_j(\sigma_j*i)$. If $lth(\sigma)$ is a limit ordinal, then $RI(\{T_j\})(\sigma) = \cup RI(\{T_j\})(\tau)$, $\tau < \sigma$.

<u>9.12 LEMMA</u> Suppose that K is α-finite and $\{T_j\}_{j\varepsilon\delta}$ is a nested sequence of trees such that if $j = j'+1$, then T_j is obtained from $T_{j'}$ by applying either the function Sp or Fu with respect to some reduction procedure e, while if j is a limit ordinal, then $T_j = RI(\{T_{j'}\})$, for $j' < j$ and $j' \varepsilon K$. Then $U = RI(\{T_j\})$ is a (nonempty) tree with the property that every string on the tree having a proper extension on T_j has one on the tree as well.

Proof. First of all, since the only operations used in forming subtrees in the sequence are Sp and Fu, we have $T_{j'}(\emptyset) < T_j(\emptyset)$ for $j' < j$. Suppose that τ lies on U and $\tau = T_j(\sigma_j)$ for some σ_j. Since for $j' < j$, T_j is a subtree of $T_{j'}$, and since trees map incompatible strings to incompatible strings, we must have $\{\sigma_j\}$ to be a sequence of compatible strings. Now suppose that $j = j'+1$. Suppose that for some e and some $\rho_{j'}$ on $T_{j'}$, we have $T_j = Sp(T_{j'},e,\rho_{j'})$. Then σ is compatible with $\rho_{j'}$ (it is in fact an extension of $\rho_{j'}$). Then by the definition of a splitting tree, we have $T_j(\sigma_j*i) \supset T_{j'}(\sigma_{j'}*i)$. If on the other hand $T_j = Fu(T_{j'},\rho_{j'})$, then clearly we have $T_j(\sigma_j*i) = T_{j'}(\sigma_{j'}*i)$. It follows that if j is a limit ordinal, then we have $\{T_{j'}(\sigma_{j'}*i)\}_{j'<j}$ is a seqeunce of compatible strings whose union belongs to T_j and properly extends τ. Using induction one may then conclude that there is a string on U which properly extends τ.

Let C be a complete α-r.e. regular set. We again denote by $\sigma lcf(\alpha,C)$ the $\Sigma_1(L_\alpha,C)$ cofinality of α, and by $\sigma lp(\alpha,C)$ the $\Sigma_1(L_\alpha,C)$ projectum of α. The latter is sometimes denoted $\alpha^*(C)$. The reader may refer to Chapter 1 for more details. Recall that $\sigma lcf(\alpha,C)$ is also the least ordinal γ for which there exists a function weakly α-recursive in C mapping γ cofinally into α. It is not difficult to see that C is hyperregular if and only if $\sigma lcf(\alpha,C) = \alpha$, if and only if every α-r.e. set is regular and hyperregular, and if and only if α is Σ_2-admissible. Recall also that $\alpha^*(C)$ is the least ordinal γ for which every $\Sigma_1(L_\alpha,C)$ set bounded below γ is $\langle L_\alpha,C\rangle$ finite (i.e. an element of $\langle L_\alpha,C\rangle$). For C regular (as is true in this case) and $\sigma lcf(\alpha,A) \geq \sigma lp(\alpha,A)$, we have $\alpha^*(C)$ = the least ordinal γ for which there exists a $\Sigma_1(L_\alpha,C)$ injection from α into γ (hence a $\Sigma_1(L_\alpha,C)$ partial function from γ onto α).

Let X be a subset of α. From now on we will write α-X-finite, α-X-recursive, α-X-r.e. etc. for $\langle L_\alpha,X\rangle$-finite, $\Delta_1(\langle L_\alpha,X\rangle)$, and $\Sigma_1(\langle L_\alpha,X\rangle)$ sets. We leave it to the reader to verify that $\sigma lp(\alpha,C) = \sigma 2p(\alpha)$. Finally note that a partial α-C-recursive map is a function whose graph is α-C-r.e. The next lemma is virtually a restatement of Lemma 8.19.

9.13 LEMMA Suppose that $\sigma lcf(\alpha,C) > \alpha^*(C)$. There is a total α-C-recursive map p from $\sigma lcf(\alpha,C)$ onto α. Furthermore, for all $\gamma < \sigma lcf(\alpha,C)$, $p|\gamma$ is an α-C-finite function.

Proof. If $\sigma lcf(\alpha,C) = \alpha$, let p be the identity function. Suppose that $\sigma lcf(\alpha,C) < \alpha$. Let $p_2: \sigma 2p(\alpha) \to \alpha$ be an α-C-recursive projection of a subset of $\sigma 2p(\alpha)$ ($= \alpha^*(C)$) onto α. Let f: $\sigma lcf(\alpha,C) \to \alpha$ be an α-C-cofinality function. For $x < \alpha^*(C)$ and $\gamma < \sigma lcf(\alpha,C)$ let $p(\langle x,\gamma\rangle) = z$ if and only if $p_2(x) = z$ and this computations converges in less than $f(\gamma)$ steps. Clearly p is α-C-recursive and onto. One obtains from this a bijection between $\sigma lcf(\alpha,C)$ and α. The second part of the assertion follows from the fact that the domain of p is equal to $\sigma lcf(\alpha,C)$.

Recall that an α-C-cardinal is a cardinal in the sense of $\langle L_\alpha,C\rangle$.

The next lemma is a relativized version of Lemma 4.1 (the Sacks-Simpson Lemma).

9.14 LEMMA Let $\langle L_\delta,T\rangle$ be an admissible structure. Suppose that κ is a regular δ-T-cardinal. Let K be a δ-T-finite set of δ-T-cardinality less than κ. If $\{I_j | j \in K\}$ is a uniformly δ-T-r.e. sequence of δ-T-finite sets each of δ-T-cardinality less than κ, then $\bigcup I_j$, j in K, is δ-T-finite and of δ-T-cardinality less than κ.

Proof. The proof follows closely that of Lemma 4.1. Let f be a one-one δ-T-recursive enumeration of $X = \bigcup I_j \times \{j\}$, j in K. If the domain of f is less than κ, then by the fact that $\kappa < \delta$, X must be δ-T-finite. Suppose that the domain contains κ. Consider $f''\kappa$. Again since $\kappa < \delta$, and f is δ-T-recursive, $f''\kappa$ is δ-T-finite and of δ-T-cardinality κ. Let $J_j = I_j \times \{j\} \cap f''\kappa$. Then each J_j has δ-T-cardinality less than κ, and $f''\kappa = \bigcup J_j$, j in K. Since K has δ-T-cardinality less than κ, and the union is δ-T-recursive, this contradicts the regularity of κ.

<u>9.15 LEMMA</u> Let $\delta = \sigma 1cf(\alpha,C)$. There is a set $T \subseteq \delta$ such that

(a) T is $\Delta_1(\langle L_\alpha,C\rangle)$.

(b) T is δ-regular.

(c) $\langle L_\delta,T\rangle$ is admissible (i.e. $\sigma 1cf(\delta,T) = \delta$).

(d) For all $A \subseteq \delta$, A is $\Delta_1(\langle L_\delta,T\rangle)$ (resp. $\Sigma_1(\langle L_\delta,T\rangle)$) if and only if it is $\Delta_1(\langle L_\alpha,C\rangle)$ (resp. $\Sigma_1(\langle L_\alpha,C\rangle)$).

Proof. If $\delta = \alpha$, then we may take T to be C. Suppose $\delta < \alpha$. Let p be as in Lemma 9.13. Let $\{(\exists y)(\psi_{p(\gamma)}(x,y)\}$ be an α-C-recursive list of all α-C-recursive predicates. Define

$$T = \{\langle x,y,\gamma\rangle \mid \langle L_\alpha,C\rangle \models \psi_{p(\gamma)}(p(x),p(y)\}.$$

Observe that T is a subset of δ, and T is α-C-recursive. Now for all $\gamma < \delta$, $T|\gamma$ is α-finite (by the fact that p is α-C-recursive and therefore $\Sigma_2(L_\alpha)$, and that $\gamma < \delta = \sigma 2cf(\alpha)$). Since $\delta < \alpha$, δ is an α-C-cardinal (hence α-cardinal), and so is α-stable (see remark following Corollary 4.4). As $T|\gamma$ is bounded below δ, it is δ-finite as well. This shows that T is δ-regular.

Now if X is δ-T-r.e., we have

$$X = \{y \mid \langle L_\delta,T\rangle \models (\exists x)\psi(x,y)\},$$

where ψ is $\Sigma_1(\langle L_\delta,T\rangle)$. But then

$$X = \{y \mid \langle L_\alpha,C\rangle \models (\exists \eta<\delta)(\exists K,H)(K \subseteq T \ \&\ H \subseteq \delta - T \ \&\ K \cup H = \eta \ \&\ \langle L_\eta,K\rangle \models (\exists x)\psi(x,y)\}.$$

Since T is α-C-recursive and $\delta < \alpha$, we have X to be α-C-r.e.

Now suppose that X is α-C-r.e. Then

$$X = \{y \mid \langle L_\alpha, C\rangle \models (\exists x)\psi_{p(\gamma)}(x,y)\}$$

for some γ. But then $X = \{y \mid (\exists x)(\langle x,y,\gamma\rangle \in T\}$. So X is δ-T-r.e.

Let $t: \gamma \to \delta$ be a δ-T-recursive map from $\gamma < \delta$ into δ. If $t''\gamma$ is unbounded in δ, then $s_\nu = \max \{p(x) \mid x < t(\nu)\}$ (where p is as in Lemma 9.13), for $\nu < \gamma$, is unbounded in α. The latter is α-C-recursive (by the above). Since $\gamma < \delta = \sigma 1cf(\alpha, C)$, we must have $\{s_\nu\}$ to be bounded in α, so that $t''\gamma$ is bounded in γ. It follows that $\langle L_\delta, T\rangle$ is an admissible structure.

The structure $\langle L_\delta, T\rangle$ in the above lemma is called the 'admissible collapse' of the structure $\langle L_\alpha, C\rangle$. It is due to Maass ([1978a]). When $\delta = \alpha$, one may simply take T to be C.

From now on we fix $\delta = \sigma 1cf(\alpha, C)$. Let $f: \delta \to \alpha$ be a $\Sigma_1(\langle L_\alpha, C\rangle)$-cofinality function. We may also assume that f is continuous; in other words, if λ is a limit ordinal then $f(\lambda) = \sup f(\delta)$, $\delta < \lambda$. Let $t\sigma 2p(\delta, T)$ be the tame Σ_2-projectum of $\langle L_\delta, T\rangle)$ (cf. Chapter 1). Let h' be δ-T-recursive such that $h': \delta \times t\sigma 2p(\delta, T) \to \delta$ and $h(j) = \lim_{\gamma \to \delta} h'(\gamma, j)$ is a tame $\Sigma_2(\langle L_\delta, T\rangle)$ projection. This means that every $j < t\sigma 2p(\delta, T)$ there is a stage γ_j where for all $\gamma \geq \gamma_j$, for all $j' < j$, we have $h'(\gamma, j') = h'(\gamma_j, j')$. Let $p: \delta \to \alpha$ be the α-C-recursive projection given in Lemma 9.13. Then $p \circ h'$ is α-C-recursive mapping $t\sigma 2p(\delta, T)$ into α. Denote this by g'. We then have g' to be a tame approximation of $g = p \circ h$, a tame $\Sigma_2(\langle L_\alpha, C\rangle)$ projection of $t\sigma 2p(\delta, T)$ onto α. We are now ready to state and prove the main result of this chapter.

<u>9.16 THEOREM</u> Suppose that $\sigma 2cf(\alpha) \geq \sigma 2p(\alpha)$. Then there exists a minimal α-degree.

Proof. We fix $\delta = \sigma lcf(\alpha,C)$ and let f, g denote the functions described above. The construction proceeds in δ many stages. We will construct a set G to be the union of G^{γ} of compatible strings such that $lth(G^{\gamma}) \geq f(\gamma)$. This ensures that G is α-regular.

At any stage $\gamma < \delta$, we have an α-C-finite sequence of trees $T_{j,\gamma}$ for $j < t\sigma 2p(\delta,T)$. Set $T_{j,0} = \emptyset$ for all $j < t\sigma 2p(\delta,T)$. Set $G^0 = \emptyset$. Suppose that we are at stage $\gamma+1$. Assume that G^{γ} lies on each $T_{j,\gamma}$ and $\{T_j\}$ is a nested sequence of trees obtained using the operations of Sp, Fu, or RI. Find the least ordinal j such that either $g'(\gamma+1,j) \neq g'(\gamma,j)$ or there does not exist incompatible strings σ_0, σ_1 extending G^{γ} lying on $T_{j,\gamma}$ such that

(9.17) if $T_{j,\gamma}(\tau) = G^{\gamma}$, then $\sigma_i \supset T_{j,\gamma}(\tau * i)$;

(9.18) $lth(\sigma_i) \geq f(\gamma+1)$.

If such a j does not exist, let $T_{j,\gamma+1} = T_{j,\gamma}$ for all j and let $G^{\gamma+1} = G^{\gamma}$. Suppose such a j exists. Let $j(\gamma+1)$ be the least such j. We consider two cases.

Case 1. $g'(\gamma,j(\gamma+1)) \neq g'(\gamma+1,j(\gamma+1))$.

For $j < j(\gamma+1)$, let $T_{j,\gamma+1} = T_{j,\gamma}$. If $j(\gamma+1)$ is a limit ordinal, compute $U = RI(\{T_{j,\gamma+1}\})$ for $j < j(\gamma+1)$. Since the operation RI is taken on a nested sequence of trees of length $j(\gamma+1) < \delta = \sigma lcf(\alpha,C)$, U satisfies our definition of a tree. Then again G^{γ} lies on U. Let $T_{j(\gamma+1),\gamma+1} = Sp(U,g'(\gamma+1,j(\gamma+1)),G^{\gamma})$. For $j > j(\gamma+1)$, set $T_{j,\gamma+1} = Sp(T_{j',\gamma+1},g'(\gamma+1,j),G^{\gamma})$ if $j = j'+1$. Otherwise we take $U(j) = RI(\{T_{j',\gamma+1}\})$, for $j' < j$, and choose $T_{j,\gamma+1}$ to be the tree $Sp(U(j),g'(\gamma+1,j),G^{\gamma})$. If $j(\gamma+1)$ is a successor ordinal, say $j'+1$, set $T_{j'',\gamma+1} = T_{j'',\gamma}$ if $j'' \leq j'$. Set

$$T_{j(\gamma+1),\gamma+1} = Sp(T_{j',\gamma+1},g'(\gamma+1,j(\gamma+1)),G^{\gamma}),$$

and define $T_{j,\gamma+1}$, for $j > j(\gamma+1)$ just as when $j(\gamma+1)$ is a limit ordinal. We set $G^{\gamma+1} = G^{\gamma}$ in either case.

Case 2. (9.17) or (9.18) fails.

Set $T_{j,\gamma+1} = T_{j,\gamma}$ for $j < j(\gamma+1)$. Suppose that $j(\gamma+1)$ is a limit ordinal. Define U as in Case 1. By induction, $T_{j(\gamma+1),\gamma} = V = Sp(U,g'(\gamma+1,j(\gamma+1)),G^{\gamma})$ and there are no strings σ_0, σ_1 extending $G^{\gamma} = V(\tau)$ satisfying (9.17) and (9.18). Suppose (9.17) fails. This means that (see Definition 9.11 and the discussion following it) there is an i such that either (a) for some x and for all strings ρ on U extending $V(\tau * i)$, the computation $\{g'(\gamma+1,j(\gamma+1))\}^{\rho}(x)$ is not defined, or (b) no incompatible extensions of $V(\tau * i)$ in U split $V(\tau)$ for $g'(\gamma+1,j(\gamma+1))$. Using C pick the least string on U extending $V(\tau * i)$ as described in either (a) or (b) and of length at least $f(\gamma+1)$. Let $G^{\gamma+1}$ be this string. Let $T_{j(\gamma+1),\gamma+1} = Fu(V,G^{\gamma+1})$. Now define $T_{j,\gamma+1}$, for $j > j(\gamma+1)$, in a manner similar to that in Case 1. Notice that $G^{\gamma+1}$ exists by our assumption on the choice of $j(\gamma+1)$, and by Lemma 9.12. Now suppose that (9.18) fails. Then there is an i such that there is no string lying on $Sp(U,g'(\gamma+1,j(\gamma+1)),V(\tau))$ extending $V(\tau * i)$ of length at least $f(\gamma+1)$. Pick the least such i. Let $G^{\gamma+1}$ be the least string on U extending $V(\tau * i)$ of length at least $f(\gamma+1)$. Observe that no incompatible strings on U extending $V(\tau * i)$ split $V(\tau * i)$ for e. Let $T_{j(\gamma+1),\gamma+1} = Fu(V,G^{\gamma+1})$. Define, for $j > j(\gamma+1)$, $T_{j,\gamma+1}$ as in the previous cases.

Now suppose that $j(\gamma+1)$ is a successor ordinal $j'+1$. Define the trees $T_{j,\gamma+1}$ as in the case when $j(\gamma+1)$ is a limit ordinal, except that now U is set to be simply $T_{j',\gamma+1}$. This completes the construction at successor stage $\gamma+1$.

Consider now the case when γ is a limit ordinal λ. Set $G^{<\lambda} = \cup G^{\gamma}$, $\gamma < \lambda$. For each j, compute $\lim_{\gamma\to\lambda} T_{j,\gamma}$. If the limit exists set it to be $T_{j,<\lambda}$. Observe that by the definition of trees and by induction, if G^γ lies on $T_{j,<\lambda}$ for each $\gamma < \lambda$, then G^λ lies on it as well. Find the least j such that either $g'(\lambda,j) \neq \lim_{\gamma\to\lambda} g'(\gamma,j)$, or $T_{j,<\lambda}$ does not exist, or there does not exist σ_0, σ_1 on $T_{j,<\lambda}$ extending $T_{j,<\lambda}(\tau) = G^{<\lambda}$ satisfying (9.17) and (9.18) (where G^γ is replaced by $G^{<\lambda}$). Set the least such j to be $j(\lambda)$. We consider each situation separately.

First of all, suppose that $\lim_{\gamma\to\lambda} g'(\gamma,j(\lambda)) = g'(\lambda,j(\lambda))$. Then upon replacing $j(\lambda)$ by $j(\gamma+1)$ and $<\lambda$ by γ we are in Case 1 above and so one may define $T_{j,\lambda}$, G^λ in an analogous way.

Now suppose that $T_{j(\lambda),<\lambda}$ does not exist. Set $G^\lambda = G^{<\lambda}$. Let $T_{j,\lambda} = T_{j,<\lambda}$ if $j < j(\lambda)$. Inductively, let $T_{j+1,\lambda} = Sp(T_{j,\lambda},g'(\lambda,j),G^\lambda)$ for $j > j(\lambda)$. If $j > j(\lambda)$ is a limit ordinal, let $T_{j,\lambda} = Sp(RI\{T_{j',\lambda}\},g'(\lambda,j),G^\lambda)$, where $j' < j$. If $j(\lambda) = j'+1$, let $T_{j(\lambda),\lambda} = Sp(T_{j',\lambda},g'(\lambda,j(\lambda)),G^\lambda)$. If $j(\lambda)$ is a limit ordinal, let $T_{j(\lambda),\lambda} = Sp((RI\{T_{j,\lambda}\},g'(\lambda,j(\lambda)),G^\lambda)$, where $j < j(\lambda)$.

Finally suppose that there do not exist σ_0, σ_1 on $T_{j(\lambda),<\lambda}$ extending $T_{j(\lambda),<\lambda}(\tau) = G^{<\lambda}$ satisfying (9.17) and (9.18). Replacing, for $j < j(\lambda)$, $T_{j,\gamma}$ by $T_{j,<\lambda}$ in Case 2 above, and G^γ by $G^{<\lambda}$ we may define our trees $T_{j,\lambda}$ (for all j) and G^λ in a similar way. This completes the construction at limit stage λ.

For each $j < t\sigma 2p(\delta,T)$, let $I_j = \{\gamma | j(\gamma) = j\}$. The collection $\{I_j\}$ is uniformly α-C-r.e., hence γ-T-r.e. by Lemma 9.15. I_j is the set of stages γ where the 'jth requirement' is injured. The following is a δ-T-finite injury lemma which leads to the minimality of G.

<u>9.19 LEMMA</u> For all $j < t\sigma 2p(\delta,T)$, I_j is δ-T-finite.

Proof. Fix $j_0 < t\sigma 2p(\delta,T)$. Since g is tame $\Sigma_2(<L_\delta,T>)$, there is a stage γ^* where $g'(\gamma,j') = g(j')$ for all $j' < j_0$ and $\gamma > \gamma^*$. We establish the lemma by induction. From now on all ordinals considered are greater than γ^*.

Suppose $\gamma_0 \supset I_{j'}$ for all $j' < j_0$. Then there can be at most two stages γ after γ_0 where $T_{j,\gamma}$ is redefined. One occurs when $T_{j_0,<\gamma}$ is not defined, and the other when either (9.17) or (9.18) fails to hold. After these two problems have been settled, say at stage $\gamma(j_0) > \gamma_0$, $T_{j_0,\gamma} = T_{j_0,\gamma(j)}$ for all stages $\gamma > \gamma(j_0)$. This is true because only the operations of Sp, Fu and RI are involved. More specifically, after stage $\gamma(j_0)$, $T_{j_0,\gamma}$ is either always the splitting tree of some tree, so that it is never redefined, or is the full subtree of some tree U (= $T_{j',\gamma}$ for some $j' < j_0$ or = $RI(\{T_{j',\gamma}\})$ where $j' < j_0$), which again is never redefined since by induction U itself is not redefined. It follows that $I_{j'} \subset \gamma(j_0)$ for all $j' < j_0$.

Suppose now that for all $j' < j_0$ there is a $\gamma(j')$ such that $\gamma(j') \supset I_{j''}$, $j'' \leq j'$. We show that $I = \bigcup I_{j'}$, $j' < j_0$, is contained in some γ_1. Now I is uniformly δ-T-r.e. We consider two cases.

Case 1. There is a δ-T-cardinal $\kappa \geq t\sigma 2p(\delta,T)$. Then $j_0 < \kappa$ and we may assume that κ is a regular δ-T-cardinal (otherwise κ is singular and we may again take some regular cardinal between κ and j_0). Suppose by induction that each $I_{j'}$, $j' < j_0$, has δ-T-cardinality less than κ. By Lemma 9.14 we conclude that I is δ-T-finite and has δ-T-cardinality less than κ as well.

Case 2. There is no δ-T-cardinal $\geq t\sigma 2p(\delta,T)$. There are two possibilities.

(a) $\sigma 2p(\delta,T) = t\sigma 2p(\delta,T)$. Let

$$R = \{(j',i) \mid j'<j_0 \ \& \ (\exists \gamma)(\forall \nu \geq \gamma)[j(\nu) \geq j' \ \& \ |\{\mu \mid \mu \geq \gamma \ \& \ j(\mu) = j_0\}| = i\}.$$

By an earlier observation, for each (j',i) in R, $i \leq 2$. Thus R is a

$\Sigma_2(<L_\delta,T>)$ set bounded below $\sigma 2p(\delta,T)$, hence δ-T-finite. For $j' < j_0$, let $t(j')$ be the stage μ where $\mu > x = \sup t(j")$, $j" < j'$ and μ is the ith time after x when $j(\mu) = j'$. Then t is δ-T-recursive and defined on all ordinals less than $j_0 < \sigma 2p(\delta,T) = t\sigma 2p(\delta,T) \leq \delta$, so that by $<L_\delta,T>$-admissibility, the range of t is δ-T-finite bounded by an ordinal γ_1. Thus $\gamma_1 \supset I$.

(b) $t\sigma 2p(\delta,T) > \sigma 2p(\delta,T)$. By the remark following Theorem 1.60 we must have $t\sigma 2p(\delta,T) = \sigma 2p(\delta,T).\sigma 2cf(\delta,T)$. If $j = \sigma 2p(\delta,T).\eta+j^*$, where $\eta < \sigma 2cf(\delta,T)$ and $0 < j^* < \sigma 2p(\delta,T)$, then again by Lemma 9.14 and induction we have I to be contained in some $\gamma_1 < \delta$. Suppose that $j^* = 0$ and $\eta = \eta'+1$. Let $\gamma' = \gamma(\sigma 2p(\delta,T).\eta')$. Suppose for each $j' < j$ such that $j' = \sigma 2p(\delta,T).\eta' + (j')^*$, the order type of $I_{j'} - \gamma'$ is less than the next δ-T-cardinal after $(j')^*$. For convenience let us set $\mu = \sigma 2p(\delta,T)$. Let

$$R = \{<x,(j')^*> \mid (\exists\gamma)(\gamma \in I' = I_{\mu.\eta+(j')^*} - \gamma(\mu.\eta) \ \& \ \text{order type } I' \text{ is } x\}.$$

Then $R \subset \mu.\mu$, is δ-T-r.e. and so δ-T-finite (we have $\sigma 1p(\delta,T) = \delta$, else it is the greatest δ-T-cardinal $\geq t\sigma 2p(\delta,T)$, contrary to the assumption that we are in Case 2). Using R and the admissibility of $<L_\delta,T>$ we conclude that I is contained in some γ_1.

Finally suppose that $\eta = \lambda$ is a limit ordinal. Then by the fact that $\lambda < \sigma 2cf(\delta,T)$, there must be a bound γ_1 on I below δ. This proves Lemma 9.19.

<u>9.20 LEMMA</u> G is of minimal α-degree.

Proof. By Lemma 9.8, G is not α-recursive. Suppose that $\{e\}^G$ is the characteristic function of a set. Let $g(j) = e$. By Lemma 9.19, there is a $\gamma(j) \supset I(_j)$. For any stage $\gamma > \gamma(j)$, $T_{j,\gamma}$ is fixed. By construction, it is either a splitting tree for e or a full tree (of a tree) for e. Furthermore, G is a branch of $T_{j,\gamma}$. By Lemmas 9.3 and 9.7, we conclude that either $\{e\}^G \geq_\alpha G$, or $\{e\}^G$ is α-recursive.

The proof of Theorem 9.16 is complete.

9.21 REMARK The existence of a minimal α-degree for $\sigma 2cf(\alpha) < \sigma 2p(\alpha)$ is still unknown. An important test case is when $\alpha = \aleph_\omega^L$. Note that $\langle L_\alpha, C\rangle$ (where C is a complete α-r.e. set) is now an inadmissible structure. In fact $\sigma 1cf(\alpha, C) < \sigma 1p(\alpha, C) = \alpha$. This is an example of what is called a 'strongly inadmissible structure' in β-recursion theory (Friedman [1976]; Maass [1978a]). The attempt to construct minimal α-degrees (for this α) was one of the motivating forces behind the investigations into recursion theory on inadmissible ordinals. Despite impressive advances made on the subject itself, and a strong evidence of the applicability of β-recursion theory to α-recursion theory (see the next chapter), the basic question of the existence of minimal α-degrees remains unsolved. In Chong [1979] it is shown that for any admissible ordinal α, a set G of minimal α-degree below **0'** must be 'recoverable' using a variant of the standard tree construction. This shows that theoretically one does not have to look too far for a construction of such a set G. How G could actually be constructed, however, is perhaps one of the most fundamental problems in α-recursion theory.

CHAPTER 10

SET-THEORETIC METHODS

Set-theoretic techniques have been successfully exploited recently to study degree theory on Σ_2 inadmissible ordinals such as $\aleph_\omega^L$ or $\aleph_{\omega_1}^L$. Using the method of Skolem hull, S. Friedman [1981b] obtained a 'natural' solution to Post's problem for all ordinals α which are limits of α-cardinals. 'Natural' here is used in the sense that the two r.e. sets of incomparable degrees are naturally defined without the need for a priority argument, something which has not been found in classical recursion theory. On the other hand, by a careful analysis of the fine structure of L in the style of Jensen [1972], he also derived a negative solution of generalized Post's problem for all singular cardinals of uncountable cofinality (under the assumption of the Generalized Continuum Hypothesis). These developments are both pleasing and significant for two reasons: Firstly it opens up a new chapter in α-recursion theory, in that the priority method, inherited from classical recursion theory, takes a back seat (its importance notwithstanding) while completely new techniques are introduced. As these techniques lack counterparts in ω-recursion theory, they also point strongly to the independent richness of recursion theory on admissible ordinals. Secondly, the set-theoretic ideas used in the study of recursion-theoretic problems went one step beyond what had been done earlier, where only basic objects such as the Σ_n- projectum and Σ_n-cofinality were used. It indicates that there is potentially a vast source of ideas and methods in other parts of mathematical logic which could be useful to the study of higher recursion theory (for more of this, see also Friedman [1980] where the diamond principle was used to study Post's problem on inadmissible ordinals). This is also an instance where a 'fusion' of several subjects in mathematical logic takes place, in line with the general development of mathematics in the past few decades.

In this chapter we discuss two topics: the natural solution of Post's problem and the negative solution to the (generalized) Post's problem for $\aleph_{\omega_1}^L$ assuming V = L. We begin by fixing some notation.

Let α be a limit of α-cardinals. For each regular α-cardinal $\kappa < \alpha$, let $S(\kappa)$ be the set of all ordinals $\gamma < \alpha$ such that the L_α cofinality of $\gamma = \kappa$. Recall that this means that there is an α-finite function mapping κ cofinally into γ. Clearly $S(\kappa)$ is an α-r.e. set.

<u>10.1 THEOREM</u> Let α be a limit of α-cardinals. Let κ_1 and κ_2 be two distinct regular α-cardinals. Then $S(\kappa_1)$ and $S(\kappa_2)$ have incomparable α-degrees.

Proof. It is sufficient to show that $S(\kappa_1)$ is not α-recursive in $S(\kappa_2)$. We prove this by contradiction. Let ϕ be a $\Delta_0(\langle L_\alpha, S(\kappa_2)\rangle$ formula such that

$$x \in \alpha - S(\kappa_1) \leftrightarrow \langle L_\alpha, S(\kappa_2)\rangle \models (\exists y)\phi(x,y).$$

We derive a contradiction by exhibiting a γ in $S(\kappa_1)$ such that

$$\langle L_\alpha, S(\kappa_2)\rangle \models (\exists y)\phi(\gamma,y).$$

Let κ be a regular α-cardinal greater than κ_1, κ_2 and all the parameters occuring in ϕ, with the property that $\langle L_\alpha, S(\kappa_2)\rangle \models \phi(\kappa^+,y)$ for some y (κ^+ is the next α-cardinal after κ). Choose a regular α-cardinal $\kappa' > \kappa$, y. For δ and limit ordinals $\lambda < \kappa_1$, let

$$H_0 = \Sigma_1 \text{ Skolem hull of } L_\kappa \cup \{\kappa^+, y\}$$

in $\langle L_{\kappa'}, S(\kappa_2)\cap L_{\kappa'}\rangle$;

$\nu_0 = H_0 \cap \kappa^+$;

$H_{\delta+1} = \Sigma_1$ Skolem hull of $L_\kappa \cup \{\kappa^+, y, \nu_\delta\}$

in $\langle L_{\kappa'}, S(\kappa_2)\ L_{\kappa'}\rangle$;

$\nu_{\delta+1} = H_{\delta+1} \cap \kappa'$;

$H_\lambda = \bigcup H_\delta$, $\delta < \lambda$, and

$\nu_\lambda = \sup \nu_\delta$, $\delta < \lambda$, if λ is a limit ordinal.

We prove a lemma.

<u>10.2 LEMMA</u> Fix $\delta < \kappa_1$. Every α-finite function with domain κ_2 whose graph is a subset of $H_{\delta+1}$ is $H_{\delta+1}$-finite (i.e. an element of $H_{\delta+1}$).

Proof. Let $f: \kappa_2 \to H_{\delta+1}$ be α-finite. By the operation of taking Skolem hulls, there is an injection p of $H_{\delta+1}$ into L_κ. This injection is $\Sigma_1(\langle H_{\delta+1}, S(\kappa_2)\cap H_{\delta+1}\rangle)$. Then $p\circ f: \kappa_2 \to \kappa$ is α-$S(\kappa_2)$-finite (i.e. $\langle L_\alpha, S(\kappa_2)\rangle$-finite), and hence α-finite since $S(\kappa_2)$ is α-r.e. Furthermore $p\circ f$ is bounded below κ since κ is a regular α-cardinal greater than κ_2. Then $p^{-1}\circ(p\circ f) = f$ is Σ_1 over the structure $\langle H_{\delta+1}, S(\kappa_2)\cap H_{\delta+1}\rangle$. Now f is bounded in $\langle L_{\kappa'}, S(\kappa_2)\cap\kappa'\rangle$ since κ' is regular, so f is also bounded in $H_{\delta+1}$ (since $H_{\delta+1}$ is a Σ_1 elementary substructure of $\langle L_{\kappa'}, S(\kappa_2)\cap\kappa'\rangle$. But then this implies that f is $H_{\delta+1}$-finite.

Let $H = \bigcup H_\delta$, $\delta < \kappa_1$. Let $\pi: H \to L_{\gamma_0}$ be the collapsing map. Then π is κ'-$S(\kappa_2)\cap\kappa'$-finite (H is bounded in κ' by the regularity of κ' and the fact that H is a κ_1-sequence). Let $\gamma_1 = H\cap\kappa^+$. Then $\pi(\kappa^+) = \gamma_1$ and $\gamma_0 \in S(\kappa_1)$.

<u>Claim</u> $\pi"S(\kappa_2)\cap H = S(\kappa_2)\cap\gamma_0$.

Let $\gamma \in S(\kappa_2)\cap H$. Let $f: \kappa_2 \to \gamma$ be cofinal. Then f is κ'-$S(\kappa_2)\cap\kappa'$-finite hence H-finite since H is a Σ_1 elementary substructure. It follows that $\pi(f)$ is a cofinal map from κ_2 into $\pi(\gamma)$. On the other hand, suppose that $\pi(\gamma) \in S(\kappa_2)$. Let $f\colon \kappa_2 \to H$ be chosen such that $\pi\circ f$ is cofinal in $\pi(\gamma)$. Since κ_1, κ_2 are regular and distinct, there is an unbounded subset of κ_2 such that $f"X$ is cofinal in $H_{\delta+1}$ for some δ (since H is a κ_1-sequence). Furthermore we have X to be κ'-$S(\kappa_2)\cap\kappa'$-finite, hence α-finite, and so κ-finite. It follows that f is α-finite as well. By Lemma 10.2, f is $H_{\delta+1}$-finite. Now $f\colon X \to H_{\delta+1}$ is cofinal in γ. To see this, observe that if f is cofinal in some $\gamma' < \gamma$, then $\pi(f)$ is cofinal in $\pi(\gamma') < \pi(\gamma)$, contrary to the assumption. This proves the claim.

We are now ready to prove the theorem. Since $\pi" S(\kappa_2)\cap H = S(\kappa_2)\cap\gamma_0$, and $\langle L_{\kappa'}, S(\kappa_2)\cap\kappa'\rangle \models \phi(\kappa^+), y)$ for some $y < \kappa'$, we have

$$\langle H, S(\kappa_2)\cap H\rangle \models \phi(\kappa^+, y), \text{ and so}$$

$$\langle L_{\gamma_0}, \pi" S(\kappa_2)\cap H\rangle \models \phi(\pi(\kappa^+), \pi(y)),$$

which gives $\langle L_{\gamma_0}, S(\kappa_2)\cap\gamma_0\rangle \models \phi(\pi(\kappa^+), \pi(y))$. Since $\pi(\kappa^+) = \gamma_1$, and $\gamma_1 \in S(\kappa_1)$, this contradicts the fact that ϕ defines the complement of $S(\kappa_1)$ in α.

<u>10.3 REMARK</u> The argument above may be modified to show that if α is a limit of α-cardinals, then no $S(\kappa)$ is α-recursive in the join of a finite number of $S(\kappa_i)$'s, where κ, κ_i are regular α-cardinals. In Chong and Friedman [1983] it is shown that for α = the ωth constructible cardinal, there is an α-degree lying above all the $S(\kappa)$'s, and that $S(\kappa_i)$, $S(\kappa_j)$ do not form a minimal pair for integers $i \neq j$.

We now turn to a negative solution of (generalized) Post's problem. From now on assume the Axiom of Constructibility $V = L$.

Fix $\alpha = \aleph^L_{\omega_1}$ (we will drop the superscript L henceforth).

<u>10.4 LEMMA</u> $\{\aleph_\delta | \delta < \omega_1\} \equiv_\alpha \emptyset'$.

Proof. Recall that α-cardinals are α-stable ordinals (cf. Chapter 4). Thus given an α-r.e. set W_e, to see if $x \in W_e$, one looks for a δ such that $\aleph_\delta > x, e$. Then $x \in W_e$ if and only if $x \in W_e^\sigma$, where $\sigma = \aleph_\delta$. The argument carries over to arbitrary α-finite sets as well. Thus $\emptyset' \leq_\alpha \{\aleph_\delta | \delta < \omega_1\}$. On the other hand, to test whether σ is an α-cardinal, one uses $\emptyset'$ to see if there is an α-finite injection of σ into a smaller ordinal. This is an α-r.e. procedure, hence α-recursive in $\emptyset'$. It follows that $\{\aleph_\delta | \delta < \omega_1\} \leq_\alpha \emptyset'$.

As there is a $\Delta_1(L_\alpha)$ bijection between L_α and α (Chapter 1, (1.4)), for each $x \in L_\alpha$, we call the ordinal corresponding to x under this bijection <u>the order of x.</u>

<u>10.5 DEFINITION</u> Let $A \subseteq \alpha$. The cutoff function f_A of A is defined as: for $\delta < \omega_1$, $f_A(\delta)$ = the order of $A | \aleph_\delta$.

The following is easy to verify.

<u>10.6 PROPOSITION</u> If $A \geq_\alpha \emptyset'$, then f_A and A have the same α-degree.

<u>10.7 DEFINITION</u> A set $X \subseteq \omega_1$ is closed and unbounded if it is unbounded in the usual sense and if the supremum of every bounded sequence in X lies in X.

<u>10.8 DEFINITION</u> A set $X \subseteq \omega_1$ is stationary if it intersects every closed and unbounded subset of ω_1.

The proof of the following lemma is omitted (cf. Devlin [1974]).

10.9 LEMMA (Fodor's Lemma) Let κ be a regular cardinal. Let $f: \kappa \to \kappa$ such that $f(\gamma) < \gamma$ on a stationary set X (in short: f is regressive on X). Then there is a stationary set $Y \subseteq X$ on which f is a constant.

10.10 LEMMA Let $A, B \subseteq \alpha$. If $X = \{\delta \mid f_A(\delta) < f_B(\delta)\}$ is stationary, and $A, B >_\alpha \emptyset'$, then $A <_\alpha B$.

Proof. For each δ, let p_δ be the least α-finite bijection between $f_B(\delta)+1$ and $\aleph_\delta$. Let $g_B(\delta)$ be the least ν such that $p_\delta(f_A(\delta)) < \aleph_\nu$. Then g_B is regressive on the limit ordinals of X, and so by Lemma 10.9 it is a constant on some stationary subset Y of X. Suppose $g_B(\delta) = \delta_0$ for $\delta \in Y$. Then $\{p_\delta(f_A(\delta)) \mid \delta \in Y\} \subseteq \aleph_{\delta_0}$ and is therefore α-finite. Now to compute $A|\delta$ choose a δ_1 in Y such that $\delta_1 > \delta$. Use B to obtain $p_\delta^{-1}(p_\delta(f_A(\delta))$ which is the order $A|\aleph_{\delta_1}$. Thus $A <_\alpha B$.

10.11 REMARK Lemma 10.10 shows in fact a stronger result, namely that if $g <_\alpha B$ and $g(\delta) > f_A(\delta)$ on a stationary set, then $B >_\alpha A$.

10.12 THEOREM The set of α-degrees above **0'** is well-ordered with successor obtained by taking the jump operator.

Proof. Suppose that $A_1 >_\alpha A_2 >_\alpha \cdots$ is an infinite descending sequence of sets of α-degrees above **0'**. By Lemma 10.10 for each n there is a closed and unbounded set $X_n \subset \omega_1$ such that $f_{A_n}(\delta) > f_{A_{n+1}}(\delta)$ for δ in X_n. We leave it to the reader to verify that $X = \bigcap X_n$, $n < \omega$, is also closed and unbounded. But then for $\delta \in X$, $f_{A_n}(\delta) > f_{A_{n+1}}(\delta)$ for all n, yielding an infinite descending chain of ordinals, which is impossible. Thus the α-degrees above **0'** are well-ordered.

Now suppose that $\emptyset' <_\alpha A <_\alpha B$. We show that $A' <_\alpha B$. Let ϕ be a Σ_1 formula such that $A' = \{x \mid \langle L_\alpha, A\rangle \models \phi(x)\}$. For each $\gamma < \omega_1$, let

$$A(\gamma) = \{x \mid \langle L_\alpha, A \cap \aleph_\gamma \rangle \models \phi(x)\}.$$

Then $\sup_\gamma f_{A(\gamma)}(\delta) < \aleph_{\delta+1}$ for each δ. Suppose that $\sup_\gamma f_{A(\gamma)}(\delta) > f_B(\delta)$ for stationary many δ (denote this set by Z). Then for each $\delta \in Z$ there is a $\gamma(\delta)$ such that $f_{A(\gamma(\delta))}(\delta) > f_B(\delta)$. The set of such $\gamma(\delta)$'s is α-finite. Now clearly the sequence $\langle A(\gamma(\delta)) \mid \delta \in Z\rangle$ is α-recursive in A, so that $g(\delta) = f_{A(\gamma(\delta))}(\delta)$ is a function α-recursive in A. By Remark 10.11 we have $B \leq_\alpha A$, contradicting the assumption that $A <_\alpha B$.

Thus $f_B(\delta) > \sup_\gamma f_{A(\gamma)}(\delta)$ on a stationary set Z. Let p_δ be the least bijection of $f_B(\delta)+1$ onto $\aleph_\delta$. For each $\delta < \omega_1$, the cofinality of δ is ω. Thus there is an $h(\delta) < \delta$ and an unbounded (hence uncountable) set $X_\delta \subset \omega_1$ such that

$$X_\delta = \{\gamma \mid p_\delta(f_{A(\gamma)}(\delta)) < \aleph_{h(\delta)}\}.$$

The map is regressive on limit ordinals and so by Lemma 10.9 there is a stationary set $Z_1 \subset Z$ on which h takes a constant value. Let

$$Y = \{\langle \delta, p_\delta(f_{A(\gamma)}(\delta))\rangle \mid \delta \in Z_1 \ \& \ \gamma \in X_\delta\}.$$

Now Y is α-finite since $\{X_\delta\}$ is an α-finite collection. To compute $A' \mid \aleph_\delta$, use B and the fact that $B \geq \emptyset'$ to obtain $p_\delta^{-1}(y)$ for each y such that $\langle \delta, y\rangle \in Y$. Then x is in $A' \mid \aleph_\delta$ if and only if x is in the union of L_ζ, where $\zeta = p_\delta^{-1}(y)$, for some $\langle \delta, y\rangle$ in Y. Hence $A' \leq_\alpha B$.

Our next objective is to identify the α-degrees $\mathbf{a} \geq \mathbf{0}'$ using fine structure theory. We will show that if $\mathbf{a}_1 = \mathbf{0}' < \mathbf{a}_2 = \mathbf{0}'' < \ldots < \mathbf{a}_\nu < \ldots$ is a complete list of α-degrees above $\mathbf{0}'$, then each $\mathbf{a}_\nu$ is the α-degree of a set $A_\nu \subset \alpha$ which is a master code of some limit ordinal $\beta > \alpha$ in a suitable sense. For the analysis of the α-degrees above $\mathbf{0}'$ it will be more convenient to work with limit ordinals β such that $\alpha \leq$

$\beta < \alpha^+$, the next cardinal after α. Unless otherwise specified, ordinals β will always assumed to be in this range. The following notion of a master code supplements those discussed in Chapter 1.

10.13 DEFINITION Let $\beta > \alpha$ be a limit ordinal. A set $C \subseteq \delta np(\beta)$ is a $\Delta_n(L_\beta)$-master code if C is $\Delta_n(L_\beta)$ and every $A \subseteq \delta np(\beta)$ is $\Sigma_1(\langle L_{\delta np(\beta)}, C\rangle)$ if and only if it is $\Sigma_n(L_\beta)$.

An immediate consequence of the above definition is that the set $A \subseteq \delta np(\beta)$ is $\Sigma_m(\langle L_{\delta np(\beta)}, C\rangle)$ if and only if it is $\Sigma_{m+n-1}(L_\beta)$. Also, a set is $\Delta_1(\langle L_{\delta np(\beta)}, C\rangle)$ if and only if it is $\Delta_n(L_\beta)$. Our first task is to characterize the conditions under which $\delta np(\beta) = \alpha$ and where $\Delta_n(L_\beta)$-master codes exist.

10.14 THEOREM Let $\beta > \alpha$ be a limit ordinal. Then there is a $\Delta_n(L_\beta)$-master code $C \subseteq \alpha$ if and only if the $\Sigma_n(L_\beta)$-cofinality of $\sigma(n-1)p(\beta)$ (written $\sigma ncf_\beta(\sigma(n-1)p(\beta))$) is ω_1.

The proof of Theorem 10.14 utilizes some new concepts and ideas which we presently introduce. Let $\beta > \alpha$. Let $C \subseteq \beta$ such that $\langle L_\beta, C\rangle$ is β-C-regular (i.e. amenable). In analogy with the notation used in Chapter 8, let $\beta^*(C)$ be the $\Sigma_1(L_\beta)$-projectum of $\langle L_\beta, C\rangle$. This, as was shown in Chapter 1, is also the least ordinal γ for which there is $\Sigma_1(\langle L_\beta, C\rangle)$ map from a subset of γ onto β. Let $a \in L_\beta$ be the least parameter used in defining such a map. Let $h\colon \omega\times\beta \to \beta$ be a Σ_1 Skolem function for L_β with parameter a. Let $\gamma < \beta^*(C)$ and let $H_\gamma = h''\omega\times\gamma$.

10.15 DEFINITION γ is β-C-pseudostable if $\gamma \not\subseteq H_\gamma$.

10.16 LEMMA Let $\gamma < \beta^*(C)$ be β-C-pseudostable. Let $\pi\colon \langle H_\gamma, \varepsilon\rangle \to \langle L_{\gamma'}, \varepsilon\rangle$ be the collapsing map. Then γ is a γ'-cardinal.

Proof. Otherwise, there is an $L_{\gamma'}$-finite injection f of γ

into a smaller ordinal. Then one obtains from $\pi^{-1}(f)$ a set of order type γ in H_γ. So in L_β there is a map (with parameters in H_γ) taking a smaller ordinal onto γ. Since H_γ is a Σ_1 elementary substructure of $\langle L_\beta, C\rangle$, we must have $\gamma \in H_\gamma$. This contradicts the fact that γ is β-C-pseudostable.

10.17 DEFINITION Let β be a limit ordinal. The structure $\langle L_\beta, C\rangle$ is collapsible if for sufficiently large β-C-pseudostable ordinals $\gamma < \beta^*(C)$, $L_{\gamma(C)} \models \gamma$ is a cardinal, where $\gamma(C)$ is the least ordinal δ for which $\pi"C \cap H_\gamma$ is definable over L_δ.

It follows that if $\gamma < \beta^*(C)$ is β-C-pseudostable, then $\pi"C \cap H_\gamma$ is always an element of $\gamma^\wedge$, the least ordinal δ such that $L_\delta \models$ "γ is not a cardinal". If $\beta^*(C) = \alpha$, the existence of such a bound allows one to study the growth rate (in terms of the cutoff function) of a set $A \subset \alpha$ which is $\Sigma_1(\langle L_\beta, C\rangle)$, since in this case a bound on $\gamma(C)$ also establishes a bound on $A|\gamma$ (we have $A|\gamma$ to be Σ_1 over $\langle L_{\gamma'}, \pi"C \cap H_\gamma\rangle$). Observe that by Lemma 10.16 we have $\gamma' < \gamma^\wedge$. The next lemma gives a sufficient condition for the existence of collapsible structures.

10.18 LEMMA Suppose λ and β are limit ordinals such that $\delta np(\lambda) = \beta$ (resp. $\sigma(n-1)p(\lambda) = \beta$) and C is a $\Delta_n(L_\lambda) \cap$ (resp. $\Sigma_{n-1}(L_\lambda)$) master code. Then $\langle L_\beta, C\rangle$ is collapsible.

Proof. Choose a Σ_n Skolem function k for L_λ with parameters involving those used to define the set C and β-C over L_λ. As C is a Δ_n master code, $K = k \cap (\omega\times\beta\times\beta)$ is $\Sigma_1(L_\beta, C)$. The parameters used in defining the set K may be chosen to be $\gamma_0 < \beta$. It follows that if $\gamma > \gamma_0$ then $(*) \cap (k"\omega\times\gamma)|\beta = h"\omega\times\gamma = H_\gamma$. Now suppose that γ had been chosen to be β-C-pseudostable. The collapsing map π_1 for $\langle k"\omega\times\gamma, \varepsilon\rangle$ to $\langle L_{\gamma"}, \varepsilon\rangle$ extends π so that $\gamma" > \gamma'$. By (*) we still have $L_{\gamma"} \models$ "γ is a cardinal". So $\gamma" < \gamma^\wedge$. Let $K_\gamma = k"\omega\times\gamma$. Now $C \cap K_\gamma$ is $\Delta_n(L_\lambda)$, hence Δ_n

over K_γ. It follows that $\pi_1"C \cap K_\gamma$, therefore $\pi"C \cap H_\gamma$ is $\Delta_n(L_{\gamma"})$. Thus $\pi"C \cap H_\gamma$ is $\gamma^\wedge$-finite.

Before we continue with the main line of development, we pause to state without proof a fine structure theory result known as Jensen's Extension Lemma (Devlin [1974, Chapter 7, Lemma 20]).

<u>10.19 LEMMA</u> Let β be a limit ordinal. If $\rho = \sigma(n-1)p(\beta)$ and $C \subset \rho$ is a $\Sigma_n(L_\beta)$ master code, then for each limit ordinal $\eta < \rho$ such that $\langle L_\eta, C|\eta\rangle$ is amenable, $C|\eta$ is the Σ_n master code of some limit ordinal λ such that $\sigma(n)p(\lambda) = \eta$.

We now prove Theorem 10.14.

Proof of Theorem 10.14. Let $\rho = \sigma(n-1)p(\beta)$. We show that if $\sigma ncf_\beta(\rho) = \omega_1$, then a $\Delta_n(L_\beta)$ master code exists. By Corollary 1.48, we have $\delta np(\beta) = \alpha$. Let $C \subseteq \rho$ be a $\Sigma_{n-1}(L_\beta)$ master code. Then $\mathbf{A} = \langle L_\rho, C\rangle$ is amenable. Since $\sigma ncf_\beta(\rho) = \omega_1$, there is a β-finite map f from ω_1 into ρ cofinally. One may choose f to be continuous with the property that for all limit ordinal $\gamma < \omega_1$, $\mathbf{A}_\gamma = \langle L_{f(\gamma)}, C \cap f(\gamma)\rangle$ is amenable. Since C is a $\Sigma_{n-1}(L_\beta)$ master code, f is ρ-C-recursive. For each $\delta < \omega_1$, let $s(\gamma,\delta)$ be the least $f(\gamma)$-C $f(\gamma)$-pseudostable ordinal after $\aleph_\delta$. Then $\lim_\gamma s(\gamma,\delta) = s(\delta)$, the least ρ-C-pseudostable ordinal after $\aleph_\delta$. Let $D = \{s(\delta,\delta) | \delta < \omega_1\}$. Then $D \subseteq \alpha$. We claim that D is a $\Delta_n(L_\beta)$ master code. First of all, since f is ρ-C-recursive, and $\{\aleph_\delta | \delta < \omega_1\}$ is ρ-C-finite, we have $D \leq_\alpha C$. Hence every $B \subseteq \alpha$ which is $\Sigma_1(\langle L_\alpha, D\rangle)$ is $\Sigma_1(\mathbf{A})$. Conversely, suppose that $B \subseteq \alpha$ is $\Sigma_1(\mathbf{A})$. By Lemmas 10.18 and 10.19 we have $C|f(\gamma)$ to be collapsible on $f(\gamma)$. This means that $\pi"H_{s(\delta,\delta)} \cap C|f(\delta)$ is $s(\delta,\delta)^\wedge$-finite. This implies that for each δ, $B|\aleph_\delta$ is definable over $L_{s(\delta,\delta)^\wedge}$. We then have

$$x \in B \leftrightarrow \mathbf{A} \models \phi(x)$$

$$\leftrightarrow \quad (\exists \delta)(\mathbf{A}_\delta \models \phi(x))$$
$$\leftrightarrow \quad \langle L_\alpha, D\rangle \models \psi(x),$$

where ϕ is a Σ_1 formula used to define B over **A**, while ψ is a Σ_1 formula used to define B over $\langle L_\alpha, D\rangle$. Hence we have $B \subset \alpha$ is $\Sigma_1(\mathbf{A})$ if and only if it is $\Sigma_1(\langle L_\alpha, D\rangle)$. Thus D is a $\Delta_1(\mathbf{A})$ master code, hence a $\Delta_n(L_\beta)$ master code. This proves Theorem 10.14.

From now on we fix C_k to be a $\Sigma_k(L_\beta)$ master code, for $k < \omega$. Let **A**, ρ, and D denote respectively the same structure, ordinal and set as in the previous lemma. The next lemma does not depend on the value of $\sigma ncf_\beta(\rho)$, where $\rho = \sigma np(\beta)$.

<u>10.20 LEMMA</u> Let $B \subseteq \alpha$ such that $f_B(\delta) \geq s(\delta)$ for stationary many δ. Then $C_n \leq_\alpha B$.

Proof. By Lemma 10.18, **A** is collapsible. Now C_n is $\Sigma_1(\mathbf{A})$, so $C_n | \aleph_\delta$ is definable over $L_{s(\delta)^\wedge}$ for sufficiently large δ (i.e. $f_{C_n}(\delta) \leq s(\delta)^\wedge+1$). Clearly the map $\delta \to s(\delta)^\wedge+1$ is bounded by the function g such that $g(\delta)$ = the least γ where $f_B(\delta)$ is not a cardinal (since $f_B(\delta) \geq s(\delta)$). As $g \leq_\alpha B$, we have by Remark 10.11 that $C_n \leq_\alpha B$.

<u>10.21 LEMMA</u> Let β be a limit ordinal. Suppose $\sigma ncf_\beta(\rho) = \omega_1$. Suppose B is a subset of α whose α-degree is at least **0'**. If $f_B(\delta) < s(\delta)$ for stationary many δ, then either B is β-finite or $D \leq_\alpha B$.

Proof. We consider two cases.

Case 1. $f_B(\delta) < s(\delta,\delta)$ for stationary many δ.

The map $\delta \to$ the least γ such that $f_B(\delta) < s(\gamma,\delta)$ is regressive on a stationary set. By Lemma 10.9 there is a δ_0 and a stationary set X such that $f_B(\delta) < s(\delta_0,\delta)$ for each δ in the set. Now

$\langle s(\delta_0,\delta) | \delta \in X\rangle$ is ρ-finite and by Remark 10.11 B is α-recursive in this sequence, and so B is ρ-finite as well. Since $\rho < \beta$, B is β-finite.

Case 2. $f_B(\delta) > s(\delta,\delta)$ for stationary many δ.

We leave it to the reader to show that $D \leq_\alpha B$.

We now turn our attention to the case when $\sigma ncf_\beta(\rho) \neq \omega_1$. Let the function s be as defined before.

<u>10.22 LEMMA</u> Suppose that β is a limit ordinal and $\sigma ncf_\beta(\rho) \neq \omega_1$ and $\sigma np(\beta) = \alpha$. Then for all B subset of α whose α-degree lies above **0'**, if $f_B(\delta) < s(\delta)$ on a stationary set, then either B is β-finite or $C_n \leq_\alpha B$.

Proof. We consider two cases.

Case 1. $\sigma ncf_\beta(\rho) = \gamma_0 < \alpha$.

Then the map $\delta \to$ the least γ such that $f_B(\delta) < s(\gamma,\delta)$ is bounded below γ_0 if $\gamma_0 > \omega_1$ (γ_0 is a regular cardinal). In any case, whether $\gamma_0 > \omega_1$ or $\gamma_0 = \omega$, there is a $\delta_0 < \gamma_0$ and stationary many δ such that $f_B(\delta) < s(\delta_0,\delta)$. Then again B is ρ-finite, hence β-finite.

Case 2. $\gamma_0 > \alpha$.

Choose the least a such that there is a $\Sigma_1(\langle L_\rho,C_{n-1}\rangle)$ function with parameter b mapping a subset of α onto ρ. Let h be a Σ_1 Skolem function for $\langle L_\rho,C_{n-1}\rangle$ with parameter b. For each δ such that $f_B(\delta) < s(\delta)$, let $g(\delta) < \aleph_\delta$ be chosen so that $f_B(\delta) = h(n_\delta,x_\delta)$ for some $n_\delta < \omega$ and $x_\delta \in \aleph_{g(\delta)}$. The function g is regressive on a stationary set and so by Lemma 10.9 there is a stationary set X and a δ_0 such that $g(\delta) = \delta_0$ for each δ in X. Then $K_1 = \{x_\delta | x_\delta \in \aleph_{\delta_0}\}$ and $K_2 = \{n_\delta | h(n_\delta,x_\delta) = f_B(\delta)$ for δ in $X\}$ are α-finite. Since $\gamma_0 > \alpha$, $h"K_1 \times K_2$ is ρ-finite, i.e. B is ρ-finite, hence β-finite.

Recall that $\{\mathbf{a}_\nu | \nu \geq 1\}$ is the set of α-degrees $\geq_\alpha \emptyset'$. Let $\alpha < \beta_1 < \beta_2 < \ldots$ be a complete list of limit ordinals such that $\sigma np(\beta) = \alpha$ for some n. We now state our main theorem.

10.23 THEOREM For all ν, $\mathbf{a}_\nu$ is the α-degree of a subset of α which is either a Σ_n master code or a Δ_n master code (for some n) of β_ν.

Proof. We prove this by induction on ν. The theorem is clearly true for $\nu = 1$. Suppose that $\nu = \nu'+1$.

Case 1. $\mathbf{a}_\nu$ is the α-degree of a Δ_n master code $D \subseteq \alpha$ of $\beta_{\nu'}$. Let C_n be a Σ_n master of $\beta_{\nu'}$. Then C_n is $\Sigma_1(\langle L_\alpha, D\rangle)$ and so $C_n \leq_\alpha D'$, since D' is the largest in α-degree among sets which are Σ_1 in $\langle L_\alpha, D\rangle$. To show $D' \leq_\alpha C_n$, we show that the relations

$$R(x) \leftrightarrow x \in L_\alpha \ \& \ x \subseteq D'$$

$$T(x) \leftrightarrow x \in L_\alpha \ \& \ x \subseteq \alpha - D'$$

are $\Sigma_{n+1}(L_{\beta_{\nu'}})$, since then by the master code property the relations are $\Sigma_1(\langle L_\alpha, C_n\rangle)$.

Fix $x \in L_\alpha$. Then there is a Σ_1 formula ϕ such that

$$\begin{aligned} x \subseteq D' &\leftrightarrow (\forall y \varepsilon x)(\langle L_\alpha, D\rangle \models \phi(y)) \\ &\leftrightarrow (\forall y \varepsilon x)(\exists \gamma)(\langle L_\gamma, D \cap \gamma\rangle \models \phi(y)) \\ &\leftrightarrow (\exists \text{ sequence } \langle x_\delta | \delta < \omega_1\rangle \ \& \ x = \textstyle\bigcup x_\delta \\ &\qquad \& \ (\forall \delta < \omega_1)(\forall y \varepsilon x_\delta)(\langle L_{\aleph_\delta}, D \cap \aleph_\delta\rangle \models \phi(y))). \end{aligned}$$

Now the function $\delta \to \aleph_\delta$ is $\Sigma_1(\langle L_\alpha, D\rangle)$, and so $R(x)$ is $\Sigma_2(\langle L_\alpha, D\rangle)$. Since D is a $\Delta_n(L_{\beta_{\nu'}})$ master code, R is $\Sigma_{n+1}(L_{\beta_{\nu'}})$.

On the other hand, for $x \in L_\alpha$, we have

$$x \subseteq \alpha - D' \leftrightarrow (\forall \gamma)(\forall y \varepsilon x)(\langle L_\gamma, D \cap \gamma\rangle \models \sim\phi(y)),$$

so that $T(x)$ is $\Pi_1(<L_\alpha,D>)$, hence $\Pi_n(L_{\beta_{\nu'}})$. Thus R and T are both $\Sigma_{n+1}(L_{\beta_{\nu'}})$, and so $\Sigma_1(<L_\alpha,C>)$. This shows that $D' \leq_\alpha C_n$.

Case 2. $\mathbf{a}_{\nu'}$ is the α-degree of a Σ_n master code C_n for $\beta_{\nu'}$. Then $C_n = D$ is also a Δ_{n+1} master code for $\beta_{\nu'}$. By Case 1 the α-degree of C_{n+1}, a Σ_{n+1} master code for $\beta_{\nu'}$, is equal to the α-degree of C_n'.

In either case we have D' to be of the same α-degree as either C_n or C_{n+1} which in turn is a Σ_1 master code for $\beta_\nu = \beta_{\nu'}+1$. This proves the theorem at successor ordinals.

Now let $\nu = \omega.\mu$, a limit ordinal. Choose n so that $\sigma np(\beta_\nu) = \alpha < \sigma(n-1)p(\beta_\nu)$.

If $\sigma ncf(\sigma(n-1)p(\beta_\nu)) = \omega_1$, then by Lemma 10.21 there is a $\Delta_n(L_{\beta_\nu})$ master code $D \subseteq \alpha$ and for all $B \subseteq \alpha$, either B is β_ν-finite or $D \leq_\alpha B$. On the other hand, if $\sigma ncf(\sigma(n-1)p(\beta_\nu)) \neq \omega_1$, then by Lemma 10.22 the $\Sigma_n(L_{\beta_\nu})$ master code C_n has the property that for all $B \geq_\alpha \emptyset'$, either B is β_ν-finite or $C_n \leq_\alpha B$. In either case (regardless of the value of $\sigma ncf(\sigma(n-1)p(\beta_\nu))$), if B is β_ν-finite then B is α-recursive in some Σ_m or Δ_m master code for some $\beta_{\nu'}$, $\nu' < \nu$, so that the α-degree of B is $\leq \mathbf{a}_\nu$. Since the α-degrees above $\mathbf{0}'$ are well-ordered, the α-degree of D (or C_n as the case may be, depending on the cofinality of the ordinal $\sigma(n-1)p(\beta_\nu)$ in L_{β_ν}) must be equal to $\mathbf{a}_\nu$. This completes the proof of Theorem 10.23.

<u>10.24 REMARK</u> Assuming the GCH, the above results (Theorems 10.12 and 10.23) may be generalized to singular cardinals of uncountable cofinality, where the degree $\mathbf{0}'$ is replaced by an appropriate degree $\mathbf{a}$. For a survey of applications of set-theoretic methods in higher recursion theory, cf. Friedman [1983, 1984].

<u>10.25 REMARK</u> Much less is known about the α-degrees above $\mathbf{0}'$

for α a Σ_2 inadmissible constructible cardinal of countable cofinality. It is known that there exist two incomparable $\aleph_\omega^L$ degrees above $\mathbf{0}'$, but they are not definable over $\aleph_\omega^L$ (Friedman [1983]).

10.26 REMARK Yang [1984] has shown that for all admissible ordinals $\alpha > \omega$, the theory of $\langle \mathbf{D}_\alpha, < \rangle$ is not elementarily equivalent to the theory of $\langle \mathbf{D}_\omega, < \rangle$, where $\mathbf{D}_\alpha$ and $\mathbf{D}_\omega$ are respectively the structure of α-degrees and ω-degrees. Shore [1978a] has shown that the theory of α-r.e. degrees is not elementarily equivalent to the theory of ω-r.e. degrees whenever α is a Σ_2 inadmissible constructible cardinal.

APPENDIX

NEGATIVE SOLUTION TO POST'S PROBLEM

It is fair to say that the study of Post's problem, in the general setting of recursion-theoretic structures, provided a powerful thrust to the impressive development of higher recursion theory in the past thirty years. Many would even argue that it is the principal guiding force. In each instance where a recursion theory on a structure is introduced, the question of the existence of two incomparable r.e. sets over the structure naturally arises. The implications of a positive solution (if exists) are far reaching. On the one hand the technique used in the construction would provide a deeper insight into the inherent properties of the structure under study. The use or the absence of the priority method in the proof becomes especially significant. On the other hand, the result would lead immediately to a long list of questions related to the structure theory of r.e. degrees, from splitting, to density and beyond.

Thus the viability of a particular notion of reducibility depends on whether, based on that notion, Post's problem has a positive solution. It was for this reason that the relation 'hyperarithmetical in' was rejected as a proper generalization of 'recursive in' for ω_1^{CK}-recursion theory (on the reals), since every Π_1^1 set of natural numbers is either complete or hyper-arithmetical. This eventually led to the discovery of the notion of admissible ordinals, where positive solutions to Post's problem were obtained. Indeed a positive solution to Post's problem exists on an admissible set A whenever there is a 'nice' recursive well-ordering of A. The subsequent introduction of recursion theory on inadmissible ordinals, where incomparable r.e. degrees were obtained for some of them (S. Friedman [1980]), showed in fact that the assumption of admissibility on a structure is not needed to answer Post's problem positively. As it turned out, neither is it sufficient. The first result along this line was obtained by Simpson ([1974b]), who showed that under the Axiom of Determinacy, the smallest admissible set M containing the reals R solves Post's problem negatively, i.e. every $\Sigma(M)$ set is either $\Delta(M)$ or complete. The hypothesis of determinacy was later removed by Harrington (unpublished) using a difficult forcing argument. We present here Harrington's proof following S. Friedman [1981a].

THEOREM There exists an admissible set A for which Post's problem has a negative solution.

Proof. Let C be a complete $\Sigma_1(A)$ set. We will show that for every $B \subseteq A$ which is $\Sigma_1(A)$, either B is $\Delta_1(A)$ or there is a $\Sigma_1(A)$ predicate $\phi(x,y)$ such that for all x in A

$$(1) \quad x \notin C \quad \leftrightarrow \quad (\exists y)(\phi(x,y) \ \& \ y \notin B).$$

The structure A we construct shall be a generic extension of L_{ω_2}. To do this we state without proof the following result of Jensen and Solovay:

LEMMA 1 There is a Σ_1 formula ψ such that for all $\beta < \omega_2$, and functions $f\colon \beta \to L_{\omega_2}$, there is an admissible set A_f such that

$$(2) \quad A_f \cap ORD = \omega_2$$

$$(3) \quad A_f = L_{\omega_2}[\langle w_i \mid i < \beta\rangle], \text{ where } w_i \text{ is a set of witnesses for } \psi(x),\ x \in f(i).$$

$$(4) \quad A_f \models \psi(x) \quad \leftrightarrow \quad (\exists\, i)(x \in f(i)) \quad \leftrightarrow \quad (\exists\, z \in w_i)(A_f \models \psi'(z,x)), \text{ where}$$

$\psi(x)$ is $(\exists z)\psi'(z,x)$, and ψ' is $\Delta_0(A)$.

Unless otherwise indicated, we let $\alpha = \omega_2$. For each $\beta < \alpha$, let $\psi_\beta(x)$ be $\psi(\langle\beta,x\rangle)$. The association $\psi_\beta \leftrightarrow \psi(\langle\beta,x\rangle)$ is clearly $\Delta_0(L_\alpha)$. Let $C_\beta = \langle w_i | i < \beta\rangle$.

For each Σ_1 formula $\theta(t)$, we have the following two types of requirement:

$$(I)_\beta \qquad (\forall x)[\sim\theta(x) \leftrightarrow \psi_\beta(x)].$$

The objective here is to make θ 'recursive' by proving its complement to be Σ_1. To achieve this, we introduce an operator O_β such that $O_\beta(A) \subseteq A$, and O_β is monotone, i.e. $A_1 \subset A_2$ implies that $O_\beta(A_1) \subseteq O_\beta(A_2)$. We think of O_β as a guess at $\sim\theta$.

$$(II)_\beta \quad (\forall x)[x \not\in C \leftrightarrow (\exists y)(\psi_\beta(\langle x,y\rangle) \ \& \ \sim\theta(y)].$$

In this case we ensure that if $\psi_\beta(\langle x,y\rangle)$ holds, then either y is in (the set defined by) θ or x is not in (or 'forced' to be out of) C. This would force ψ to be complete.

We introduce the following set of forcing conditions:

$$\mathbf{P} = \{\langle R,A\rangle | (\exists\beta<\alpha)[A = L_\alpha[C_\beta] \ \& \ R = \langle (X)_\gamma | X = I \text{ or } II \ \& \ \gamma < \beta\rangle\}.$$

Elements of **P** are sometimes denoted p. An ordering $\lessdot$ will be defined on **P** so that $\langle R',A'\rangle \lessdot \langle R,A\rangle$ if and only if $R \subseteq R'$, $A \subseteq A'$ and

(1) If $(I)_\beta \in R$, and $A' \models \psi_\beta(x)$, $A \not\models \psi_\beta(x)$, then $x \in O_\beta(A')$.

(2) If $(II)_\beta \in R$, $A' \models \psi_\beta(\langle x,y\rangle)$, $A \not\models \psi_\beta(\langle x,y\rangle)$, then either $A' \models \theta(y)$ or $\langle R',A'\rangle \Vdash \sim C(x)$.

There is an obvious circularity inherent in the definition of

$<$ with regard to (2) above, namely $<$ is defined in terms of $\Vdash$, which is on the other hand supposed to be defined in terms of $<$. To resolve this we introduce the following notion:

DEFINITION Let $\langle R,A\rangle$, $\langle R',A'\rangle$ be fixed. Then $\langle R',A'\rangle$ is a safe extension of $\langle R,A\rangle$, written $\langle R',A'\rangle <_s \langle R,A\rangle$ if and only if $R \subseteq R'$, $A \subseteq A'$ and

(3) If $(I)_\beta \in R$, and $A' \models \psi_\beta(x)$, $A \not\models \psi_\beta(x)$, then $x \in O_\beta(A')$.

(4) If $(II)_\beta \in R$, $A' \models \psi_\beta(\langle x,y\rangle)$ and $A \not\models \psi_\beta(\langle x,y\rangle)$, then $A' \models \theta(y)$.

If $\langle R',A'\rangle$ is a safe extension of $\langle R,A\rangle$, we also say that A' is safe over $\langle R,A\rangle$.

We now define the notion of forcing for Σ_1 and Π_1 formulas: If ϕ is Σ_1, let

(5) $\langle R,A\rangle \Vdash_s \phi(x) \quad \leftrightarrow \quad A \models \phi(x)$

(6) $\langle R,A\rangle \Vdash_s \sim\phi(x) \quad \leftrightarrow \quad A' \models \sim\phi(x)$ for all A' which is safe over $\langle R,A\rangle$.

The relation $<$ is now defined to be

$\langle R',A'\rangle < \langle R,A\rangle \quad \leftrightarrow$

(7) $R \subseteq R'$ and $A \subseteq A'$.

(8) If $(I)_\beta \in R$, $A' \models \psi_\beta(x)$ and $A \not\models \psi_\beta(x)$, then $x \in O_\beta(A')$.

(9) If $(II)_\beta \in R$, $A' \models \psi_\beta(\langle x,y\rangle)$ and $A \not\models \psi_\beta(\langle x,y\rangle)$, then either $A' \models \theta(y)$ or $\langle R',A'\rangle \Vdash_s \sim C(x)$.

(10) There is a $B \supseteq A'$ such that B is safe over $\langle R,A\rangle$.

The definition of forcing now proceeds as follows: Let ϕ be Σ_1. Then

(11) $\langle R,A\rangle \Vdash \phi(x) \quad \leftrightarrow \quad A \models \phi(x)$

(12) $\langle R,A\rangle \Vdash \sim\phi(x) \quad \leftrightarrow \quad$ for all $\langle R',A'\rangle \leqslant \langle R,A\rangle$, $\langle R',A'\rangle \nVdash \phi(x)$.

We first discuss some basic properties of this notion of forcing.

<u>LEMMA 2</u> If ϕ is either Σ_1 or Π_1, then $\langle R,A\rangle \Vdash \phi(x)$ if and only if $\langle R,A\rangle \Vdash_s \phi(x)$.

Proof. Suppose that ϕ is Σ_1. Then

$$\begin{aligned} \langle R,A\rangle \Vdash \phi(x) \quad &\leftrightarrow \quad A \models \phi(x) \\ &\leftrightarrow \quad \langle R,A\rangle \Vdash_s \phi(x). \end{aligned}$$

If ϕ is $\sim\zeta$, where ζ is Σ_1, one concludes from the fact that $\leqslant_s$ is extended by $\leqslant$ that

$$\langle R,A\rangle \Vdash \phi(x) \quad \rightarrow \langle R,A\rangle \Vdash_s \phi(x).$$

Thus suppose that $\langle R,A\rangle \Vdash_s \phi(x)$ but $\langle R,A\rangle \nVdash \phi(x)$. Then there is a pair $\langle R_1,A_1\rangle \leqslant \langle R,A\rangle$ such that $\langle R_1,A_1\rangle \Vdash \zeta(x)$. This means, by the definition of $\Vdash$, that $A_1 \models \zeta(x)$. Since Σ_1 formulas persist upwards, this implies that there is a B which is safe over $\langle R,A\rangle$ such that $B \models \zeta(x)$. We then have $\langle R,B\rangle \Vdash_s \zeta(x)$. But as $\langle R,B\rangle \leqslant_s \langle R,A\rangle$, we have $\langle R,B\rangle \Vdash_s \sim\zeta(x)$. This is a contradiction.

The proof of the next two lemmas are left to the reader.

LEMMA 3 (a) $\leq$ is transitive.

(b) $\langle \mathbf{P}, \leq \rangle$ is ω_1-closed, i.e. if $\langle R_\gamma, A_\gamma \rangle_{\gamma < \omega_1}$ is a sequence of compatible elements in $\mathbf{P}$, then $\bigcup \langle R_\gamma, A_\gamma \rangle$, $\gamma < \omega_1$, is also an element of $\mathbf{P}$.

LEMMA 4 If G is $\mathbf{P}$-generic, and ϕ is either Σ_1 or Π_1, then $L_\alpha[G] \models \phi(x)$ if and only if there is a $p \in G$ such that $p \Vdash \phi(x)$.

Thus it follows that if θ is Σ_1, then in the structure $L_\alpha[G]$ we have

(13) $x \in {\sim}\theta$ (the set defined by $\sim\theta$) $\leftrightarrow$

$(\exists \langle R,A \rangle \varepsilon G)(\langle R,A \rangle \Vdash {\sim}\theta(x)) \quad \leftrightarrow$

$(\exists \langle R,A \rangle \varepsilon G)(\forall \langle R',A' \rangle \leq \langle R,A \rangle)(\langle R',A' \rangle \not\Vdash \theta(x)) \quad \leftrightarrow$

$(\exists \langle R,A \rangle \varepsilon G)(\forall \langle R',A' \rangle \leq \langle R,A \rangle)(A' \models {\sim}\theta(x)) \quad \leftrightarrow$

$(\exists \langle R,A \rangle \varepsilon G)(\forall A' \supseteq A)[A'$ safe over $\langle R,A \rangle \rightarrow A' \models {\sim}\theta(x)]$

Given a condition $\langle R,A \rangle$, we define as result of this the operator O_β (for requirement $(I)_\beta$) on extensions of $\langle R,A \rangle$ by

(14) $O_\beta(B) = \{x \mid x \in B \;\&\; (\forall A')[A'$ safe over $\langle R,A \rangle \rightarrow A' \models {\sim}\theta(x)\}$.

Observe that $O_\beta(B)$ is defined in terms of $\langle R,A \rangle$ and θ. Let G be generic. If we set

(15) $O_\beta(G) = \bigcup \{O_\beta(A') \mid \langle R',A' \rangle \leq \langle R,A \rangle \;\&\; \langle R',A' \rangle \in G\}$,

then we have $\langle R,A \rangle$ in G implies $L_\alpha[G] \models (\forall x \varepsilon O_\beta(G))({\sim}\theta(x))$. Now the intention here is to make $\sim\theta$ equivalent to ψ_β (i.e. to make θ recursive). This can be achieved by choosing a generic G such that ψ_β defines $O_\beta(G)$, and $O_\beta(G)$ is defined by $\sim\theta$. There is however a difficulty. Namely, it may happen that for some x, x is forced to be in $\sim\theta$, but is not an element of O_β. More specifically, for any $\langle R',A'\rangle \leq \langle R,A\rangle$, there is an $\langle R'',A''\rangle \leq \langle R',A'\rangle$ such that for some x in A", $\langle R'',A''\rangle$ forces $\sim\theta(x)$, but $x \notin O_\beta(A'')$ for the O_β which is defined over $\langle R',A'\rangle$. In other words, the following situation may arise:

$$(*) \quad (\forall\langle R',A'\rangle\leq\langle R,A\rangle)(\exists\langle R'',A''\rangle\leq\langle R',A'\rangle)(\exists x\varepsilon A'')(\exists B\supseteq A'')$$
$$[B \text{ is safe over } \langle R',A'\rangle \ \& \ B\models \theta(x) \ \&$$
$$\langle R'',A''\rangle \Vdash \sim\theta(x)]$$

We will show that if $\sim(*)$ holds, then the attempt to make θ recursive will be successful. Thus suppose that $\sim(*)$ holds at $\langle R,A\rangle$. This means that

$$(16) \quad (\exists\langle R',A'\rangle\leq\langle R,A\rangle)(\forall\langle R'',A''\rangle\leq\langle R',A'\rangle)(\forall x\varepsilon A'')(\forall B)$$
$$[B \supseteq A'' \ \& \ B \text{ is safe over } \langle R',A'\rangle \ \& \ B\models \theta(x) \rightarrow$$
$$\langle R'',A''\rangle \not\Vdash \sim\theta(x)].$$

Fix such a pair $\langle R',A'\rangle$. Let $\beta < \alpha$ be an ordinal not mentioned in $\langle R',A'\rangle$, and let $(I)_\beta$ be the corresponding requirement for θ, with operator O_β defined from $\langle R',A'\rangle$ and θ. Let $R'' = R' \cup \{(I)_\beta\}$.

<u>LEMMA 5</u> $\langle R'',A'\rangle \Vdash \theta$ is recursive.

Proof. It is sufficient to show that

$$\langle R'',A'\rangle \Vdash (\forall x)(\psi_\beta(x) \leftrightarrow \sim\theta(x)).$$

Let G be generic containing $\langle R'',A'\rangle$. Let x be in $O_\beta(G)$. Then there is an $\langle R_1,A_1\rangle \ \varepsilon\ G$ such that $\langle R_1,A_1\rangle \leq \langle R',A'\rangle$ and $x \ \varepsilon\ O_\beta(A_1)$. Thus for all $B \supseteq A_1$ which is safe over $\langle R',A'\rangle$, $B\models \sim\theta(x)$.

First suppose that $L_\alpha[G] \models \theta(x)$. Then there is an $\langle R_2,A_2\rangle \ \varepsilon\ G$ such that $\langle R_2,A_2\rangle \Vdash \theta(x)$. It follows that there is an

$\langle R_2,A_2\rangle \in G$ such that $\langle R_2,A_2\rangle \leq \langle R_1,A_1\rangle$ and $\langle R_2,A_2\rangle \Vdash \theta(x)$. Since $\langle R_2,A_2\rangle \leq \langle R_1,A_1\rangle$, for all $B \supseteq A_2$ which is safe over $\langle R',A'\rangle$, $B \models \sim\theta(x)$. Also, from the fact that $\langle R_2,A_2\rangle \Vdash \theta(x)$ we have $A_2 \models \theta(x)$ and so for all $B \supseteq A_2$, $B \models \theta(x)$. This of course is a contradiction. Hence we get

$$L_\alpha[G] \models (\forall x)[(\psi_\beta(x) \leftrightarrow x \in O_\beta(G)) \;\& \; (x \in O_\beta(G) \to \sim\theta(x))].$$

To prove the converse, suppose that $L_\alpha[G] \models \sim\theta(x)$. Then there is an $\langle R_1,A_1\rangle \in G$ such that $\langle R_1,A_1\rangle \Vdash \sim\theta(x)$ with $\langle R_1,A_1\rangle \leq \langle R'',A'\rangle$, so that $\langle R_1,A_1\rangle \leq \langle R',A'\rangle$. Using $\sim(*)$, we conclude that for all $B \supset A_1$ such that B is safe over $\langle R',A'\rangle$, $B \models \sim\theta(x)$, i.e. $x \in O_\beta(A_1)$. This proves the lemma.

From now on we assume that (*) holds at $\langle R,A\rangle$. In other words, for every $\langle R',A'\rangle \leq \langle R,A\rangle$, there is an $\langle R'',A''\rangle \leq \langle R',A'\rangle$ and an x in A'', a $B \supset A''$ which is safe over $\langle R',A'\rangle$ such that $B \models \theta(x)$ and $\langle R'',A''\rangle \Vdash \sim\theta(x)$. We will show that under this assumption, θ is complete. Let $\hat{R} = R \cup \{(II)_\beta\}$.

<u>Claim 1</u> $\langle \hat{R},A\rangle \Vdash (\forall x)[\sim C(x) \leftrightarrow (\exists y)(\psi_\beta(\langle x,y\rangle) \;\& \sim\theta(y))]$.

The claim is established in a number of steps. First of all, we have

$$(17) \quad \langle \hat{R},A\rangle \Vdash (\forall x)(\forall y)[\psi_\beta(\langle x,y\rangle) \;\& \sim\theta(y) \to \sim C(x)].$$

This fact is immediate and we leave its verification to the reader. Next let $\langle R',A'\rangle \leq \langle \hat{R},A\rangle$ and $x_0 \in A'$ be chosen such that $\langle R',A'\rangle \Vdash \sim C(x_0)$. The goal now is to select an $\langle R'',A''\rangle \leq \langle R',A'\rangle$ and a $y \in A''$ such that $\langle R'',A''\rangle \Vdash \sim\theta(y)$ & $\psi_\beta(\langle x_0,y\rangle)$. There is however some difficulty in attaining this goal. Let us elaborate. If we only choose a witness w to put it into A", to make $\langle R'',A''[w]\rangle \Vdash \psi_\beta(\langle x_0,y\rangle)$, there is the problem of ensuring that $\langle R'',A''[w]\rangle \leq \langle R',A'\rangle$ and $\langle R'',A''[w]\rangle \Vdash \sim\theta(y)$. This difficulty arose because we are considering requirements of type (II) rather than type (I) whose witnesses can be easily obtained. The idea then is to transform $(II)_\beta$ into a requirement of type (I), and in the process define O_β so that forcing is preserved. This is carried out as follows:

Given $\langle T,D\rangle \leq \langle R',A'\rangle$, let T* be the set obtained from T by changing $(II)_\beta$ to $(I)_\beta$ (for θ). Let

$$O_\beta{}^*(B) = \{\langle x,y\rangle \mid \langle x,y\rangle \in B \;\&\; (B \models \theta(y) \text{ or } [x = x_0 \;\&\; \text{there exists } B' \supseteq B \text{ safe over } \langle R',A'\rangle \;\&\; B' \models \theta(y)])\}.$$

The proof of the next lemma is left to the reader.

<u>LEMMA 6</u> Let $\langle T,D\rangle \leq \langle R',A'\rangle$. Then

(a) If E is safe over $\langle T,D\rangle$, then E is safe over $\langle T^*,D\rangle$.

(b) If E is safe over $\langle R'^*,A'\rangle$, then there exists $F \supseteq E$ such that F is safe over $\langle R',A'\rangle$.

<u>COROLLARY 1</u> (a) If ϕ is Σ_1, then $\langle T,D\rangle \Vdash \phi$ if and only if $\langle T^*,D\rangle \Vdash \phi$.

(b) If ϕ is π_1, then $\langle T^*,D\rangle \Vdash \phi$ implies $\langle T,D\rangle \Vdash \phi$.

(c) If ϕ is π_1, then $\langle R',A'\rangle \Vdash \phi$ implies $\langle R'^*,A'\rangle \Vdash \phi$.

<u>COROLLARY 2</u> (*) holds at $\langle (R')^*,A'\rangle$.

Proof. Observe first of all that since (*) holds at <R,A>, it holds at all extensions of <R,A>. It follows that (*) holds at <R',A'>, and so at <(R')*,A'> as well by (a).

By (*) there is a pair $\langle (R'')^*, A''\rangle < \langle (R')^*, A'\rangle$, a y_0 in A", such that for some $D \supseteq A''$ which is safe over $\langle (R')^*, A'\rangle$, $D \models \theta(y_0)$ and $\langle (R'')^*, A''\rangle \Vdash \sim\theta(y_0)$. By Lemma 6(b) there is a $D' \supseteq D$ which is safe over $\langle R'', A''\rangle$, and since θ is Σ_1 we have $D' \models \theta(y_0)$. This implies that $\langle x_0, y_0\rangle \in O_\beta^*(A'')$. Suppose now that w is a witness to $\psi_\beta(\langle x_0, y_0)$ obtained generically (so that it will not interfere with other ψ_δ, for $\delta \neq \beta$), then we have $\langle (R'')^*, A''[w]\rangle <_s \langle (R'')^*, A''\rangle$. This implies that

$$\langle (R'')^*, A''[w]\rangle \Vdash \sim\theta(y_0) \ \& \ \psi_\beta(\langle x_0, y_0\rangle).$$

By Corollary 1, we get

$$\langle R'', A''[w]\rangle \Vdash \sim\theta(y_0) \ \& \ \psi_\beta(\langle x_0, y_0\rangle).$$

Claim 2. $\langle R'', A''[w]\rangle < \langle R', A'\rangle$.

If $\beta \neq \beta'$, one observes that the required facts hold for $\langle R''^*, A''[\omega\ ]\rangle$, and so for $\langle R'', A''[\omega]\rangle$ as well by Corollary 1. If $\beta = \beta'$, we have by assumption $\langle R', A\rangle \Vdash \sim C(x_o)$, so that $\langle R'^*, A'\rangle \Vdash \sim C(x_o)$ by Corollary 1 (c). As $\langle R''^*, A''[\omega]\rangle < \langle R'^*, A'\rangle$, we have $\langle R''^*, A''[\omega]\rangle \Vdash \sim C(x_o)$. By Corollary 1 (b), we have $\langle R'', A''[\omega]\rangle \Vdash \sim C(x_o)$. This proves Claim 2 and completes the proof of the Theorem.

BIBLIOGRAPHY

C. T. Chong [1976], An α-finite injury method of the unbounded type, J. Symbolic Logic 41, 1-17

C. T. Chong [1979], Generic sets and minimal α-degrees, Trans. Amer. Math. Soc. 254, 157-169

C. T. Chong [1979a], Major subsets of α-recursively enumerable sets, Israel J. Math. 34, 106-114

C. T. Chong [1982], Global and local admissibility, in: Proc. Patras Logic Symposium (Ed. G. Metakides), North Holland

C. T. Chong [1983], Global and local admissibility: II. Major subsets and automorphisms, Annals Pure and Applied Logic 24, 99-111

C. T. Chong and S. D. Friedman [1983], Degree theory on $\aleph_\omega^L$, Annals Pure and Applied Logic 24, 87-97

C. T. Chong and M. Lerman [1976], Hyperhypersimple α-r.e. sets, Annals Math. Logic 9, 1-48

K. J. Devlin [1974], An introduction to the fine structure of the constructible hierarchy, in Generalized Recursion Theory (Ed. J. Fenstad et. al), North Holland

K. J. Devlin [1974a], Aspects of Constructibility, Lecture Notes in Math Vol. 354, Springer Verlag

G. Driscoll [1968], Metarecursively enumerable sets and their metadegrees, J. Symbolic Logic 33, 389-411

R. Epstein and D. Posner [1978], Diagonalization in degree constructions, J. Symbolic Logic 43, 280-283

R. M. Friedberg [1957], Two recursively enumerable sets of incomparable degrees of unsolvability, Proc. Nat. Acad. Sci. (USA) 43, 236-238

R. M. Friedberg [1957a], A criterion for completeness of degrees of unsolvability, J. Symbolic Logic 22, 159-160

R. M. Friedberg [1958], Three theorems on recursive enumeration: I. Decomposition, II. Maximal set, III. Enumeration without duplication, J. Symbolic Logic 23, 309-316

S. D. Friedman [1976], Recursion on Inadmissible Ordinals, Ph.D. Thesis, Massachusetts Institute of Technology

S. D. Friedman [1978], Negative solutions to Post's problem, I. in: Generalized Recursion Theory II (Ed. J. Fenstad et. al.), North Holland

S. D,. Friedman [1979], β-recursion theory, Trans. Amer. Math. Soc. 255, 173-200

S. D. Friedman [1981], Lecture Notes in Higher Recursion Theory, Massachusetts Institute of Technology (unpublished)

S. D. Friedman [1981a], Natural α-r.e. degrees, in: Logic Year 1979 - 1980 (Ed. M. Lerman et. al.), Lecture Notes in Math. Vol. 859, Springer Verlag

S. D. Friedman [1981b], Negative solutions to Post's problem, II., Annals Math. 113, 25-43

S. D. Friedman [1983], Some recent developments in higher recursion theory, J. Symbolic Logic 48, 629-642

S. D. Friedman [1984], Fine structure theory and its applications, in: Proc. Summer Research Institute in Recursion Theory, Proc. Symposia in Pure Math., Amer. Math. Soc., to appear

S. D. Friedman and G. E. Sacks [1977], Inadmissible recursion theory, Bull. Amer. Math. Soc. 83, 255-256

R. B. Jensen [1972], The fine structure of the constructible universe, Annals Math. Logic 4, 229-308

C. G. Jockusch [1981], Degrees of generic sets, in Proc. Leeds Logic Conference (Ed. F. R. Drake et. al.), Cambridge Univ. Press

S. C. Kleene [1938], On notation for ordinal numbers, J. Symbolic Logic 3, 150-155

G. Kreisel [1971], Some reasons for generalizing recursion theory, in: Logic Colloquium '69 (Ed. C. E. M. Yates et. al.), North Holland

G. Kreisel and G. E. Sacks [1965], Metarecursive sets, J. Symbolic Logic 30, 318-336

S. Kripke [1964], Transfinite recursion on admissible ordinals, I, II (abstracts), J. Symbolic Logic 29, 161-162

A. H. Lachlan [1966], Lower bounds for pairs of recursively enumerable degrees, Proc. London Math. Soc. 16, 537-569

A. H. Lachlan [1968], On the lattice of recursively enumerable sets, Trans. Amer. Math. Soc. 130, 1-37

A. H. Lachlan [1968a], The elementary theory of recursively enumerable sets, Duke Math. J. 35, 123-146

A. H. Lachlan [1976], A recursively enumerable degree which will not split over all lesser ones, Annals. Math. Logic 9, 337-366

A. Leggett [1974], Maximal α-r.e. sets and their complements, Annals Math. Logic 6, 293-357

A. Leggett and R. A. Shore [1976], Types of simple α-recursively enumerable sets, J. Symbolic Logic 41, 681-694

M. Lerman [1972], On suborderings of the α-recursively enumerable degrees, Annals Math. Logic 4, 369-392

M. Lerman [1973], Admissible ordinals and priority arguments, in: Cambridge Summer School on Math. Logic (Ed. A. R. D. Mathias et. al.), Lecture Notes in Math Vol. 337, Springer Verlag

M. Lerman [1974], Maximal α-r.e. sets, Trans. Amer. Math. Soc. 188, 341-386

M. Lerman [1976], Types of simple α-recursively enumerable sets, J. Symbolic Logic 41, 419-426

M. Lerman [1983], Degrees of Unsolvability, Perspectives in Math. Logic, Springer Verlag

M. Lerman and G. E. Sacks [1972], Some minimal pairs of α-recursively enumerable degrees, Annals Math. Logic 4, 415-442

M. Lerman and S. G. Simpson [1973], Maximal sets in α-recursion theory, Israel J. Math. 4, 236-247

A. Levy [1965], A hierarchy of formulas in set theory, Memoirs Amer. Math. Soc. Vol. 57, 1-76

W. Maass [1977], Minimal pairs and minimal degrees in higher recursion theory, Z. Math. Logik 18, 169-186

W. Maass [1978a], Inadmissibility, tame r.e. sets and the admissible collapse, Annals Math. Logic 13, 149-170

W. Maass [1978b], The uniform regular set theorem in α-recursion theory, J. Symbolic Logic 43, 270-279

W. Maass [1978c], High α-recursively enumerable degrees, in: Generalized Recursion Theory II (Ed. J. Fenstad et. al.), North Holland

W. Maass [1978d], Fine structure of the constructible universe in α and β recursion theory, in: Higher Set Theory (Ed. G. H. Müller et. al.), Lecture Notes in Math. Vol. 669, Springer Verlag

J. W. MacIntyre [1973], Minimal α-recursion theoretic degrees, J. Symbolic Logic 38, 18-28

A. A. Muchnik [1957], On the unsolvability of the problem of reducibility in the theory of algorithms (Russian), Doklady Akad. Nauk SSSR 108, 194-197

R. Platek [1966], Foundations of Recursion Theory, Ph.D. Thesis, Stanford Univ.

W. Richter [1968], Constructively accessible ordinal numbers, J. Symbolic Logic 33, 43-55

H. Rogers Jr. [1967], Theory of Recursive Functions and Effective Computability, McGraw-Hill

G. E. Sacks [1963a], Degrees of Unsolvability, Annals Math. Studies, Princeton Univ. Press

G. E. Sacks [1963b], On the degrees less than **0'**, Annals Math. 77, 211-231

G. E. Sacks [1964], The recursively enumerable degrees are dense, Annals Math. 80, 300-312

G. E. Sacks [1966a], Post's problem, admissible ordinals and regularity, Trans. Amer. Math. Soc. 124, 1-23

G. E. Sacks [1966b], Metarecursively enumerable sets and admissible ordinals, Bull. Amer. Math. Soc. 72, 59-64

G. E. Sacks [1967], Metarecursion theory, in Sets, Models and Recursion Theory (Ed. J. Crossley), North Holland

G. E. Sacks [1971], Forcing with perfect closed sets, in Axiomatic Set Theory, Proc. Symposia in Pure Math. Vol 13, Amer. Math. Soc.

G. E. Sacks [1982], E-recursion, 1982 Summer Research Institute in Recursion Theory, Amer. Math. Soc. (mimeographed notes)

G. E. Sacks and S. G. Simpson [1972], The α-finite injury method, Annals Math. Logic 4, 323-367

J. R. Shoenfield [1972], Degrees of Unsolvability, North Holland

R. A. Shore [1972], Minimal α-degrees, Annals Math. Logic 4, 393-414

R. A. Shore [1975a], Splitting an α-recursively enumerable set, Trans. Amer. Math. Soc. 204, 65-78

R. A. Shore [1975b], The irregular and non-hyperregular α-r.e. degrees, Israel J. Math. 22, 28-41

R. A. Shore [1976], The recursively enumerable α-r.e. degrees are dense, Annals Math. Logic 9, 123-155

R. A. Shore [1976a], On the jump of an α-recursively enumerable set, Trans. Amer. Math. Soc. 217, 351-363

R. A. Shore [1977], α-recursion theory, in: Handbook of Mathematical Logic (Ed. K. J. Barwise et. al.), North Holland

R. A. Shore [1978], Some more minimal pairs of α-r.e. degrees, Z. Math. Logik und Grund. Math. 24, 409-418

R. A. Shore [1978a], On the AE-sentences of α-recursion theory, in: Generalized Recursion Theory II (Ed. J. Fenstad et. al.), North Holland

J. Silver [1974], On the singular cardinals problem, in: Proc. Int. Congress Math. (Vancouver)

S. G. Simpson [1971], Recursion Theory on Admissible Ordinals, Ph.D. Thesis, Massachusetts Institute of Technology

S. G. Simpson [1974], Degree theory on admissible ordinals, in: Generalized Recursion Theory (Ed. J. Fenstad), North Holland

S. G. Simpson [1974a], Post's problem for admissible sets, ibid.

S. G. Simpson [Notes], Introduction to admissible recursion theory (mimeographed notes)

S. G. Simpson [1978], Short course on admissible recursion theory, in : Generalized Recursion Theory II (Ed. J. Fenstad), North Holland

R. I. Soare [1976], The infinite injury priority method, J. Symbolic Logic 41, 513-530

R. I. Soare [1984], Recursively Enumerable Sets and Degrees,

Perspectives in Math. Logic, Springer Verlag

C. Spector [1956], On degrees of recursive unsolvability, Annals Math. 64, 581-592

J. Stillwell [1969], Reducibility in Generalized Recursion Theory, Ph.D. Thesis, Massachusetts Institute of Technology

J. Sukonick [1969], Lower Bounds for Pairs of Metarecursively Enumerable Degrees, Ph.D. Thesis, Massachusetts Institute of Technology

G. Takeuti [1960], On the recursive functions of ordinal numbers, J. Math. Soc. Japan 12, 119-128

G. Takeuti [1965], Recursive functions and arithmetical functions of ordinal numbers, in: Logic, Methodology and Philosophy of Science, North Holland

D. Yang [1984], On the embedding of α-recursive presentable lattices into α-recursive degrees below $\mathbf{0}'$, J. Symbolic Logic (to appear)

C. E. M. Yates [1966], A minimal pair of recursively enumerable degrees, J. Symbolic Logic 31, 159-168

INDEX

Vol. 1008: Algebraic Geometry. Proceedings, 1981. Edited by J. Dolgachev. V, 138 pages. 1983.

Vol. 1009: T. A. Chapman, Controlled Simple Homotopy Theory and Applications. III, 94 pages. 1983.

Vol. 1010: J.-E. Dies, Chaînes de Markov sur les permutations. IX, 226 pages. 1983.

Vol. 1011: J. M. Sigal. Scattering Theory for Many-Body Quantum Mechanical Systems. IV, 132 pages. 1983.

Vol. 1012: S. Kantorovitz, Spectral Theory of Banach Space Operators. V, 179 pages. 1983.

Vol. 1013: Complex Analysis – Fifth Romanian-Finnish Seminar. Part 1. Proceedings, 1981. Edited by C. Andreian Cazacu, N. Boboc, M. Jurchescu and I. Suciu. XX, 393 pages. 1983.

Vol. 1014: Complex Analysis – Fifth Romanian-Finnish Seminar. Part 2. Proceedings, 1981. Edited by C. Andreian Cazacu, N. Boboc, M. Jurchescu and I. Suciu. XX, 334 pages. 1983.

Vol. 1015: Equations différentielles et systèmes de Pfaff dans le champ complexe – II. Seminar. Edited by R. Gérard et J. P. Ramis. V, 411 pages. 1983.

Vol. 1016: Algebraic Geometry. Proceedings, 1982. Edited by M. Raynaud and T. Shioda. VIII, 528 pages. 1983.

Vol. 1017: Equadiff 82. Proceedings, 1982. Edited by H. W. Knobloch and K. Schmitt. XXIII, 666 pages. 1983.

Vol. 1018: Graph Theory, Łagów 1981. Proceedings, 1981. Edited by M. Borowiecki, J. W. Kennedy and M. M. Sysło. X, 289 pages. 1983.

Vol. 1019: Cabal Seminar 79–81. Proceedings, 1979–81. Edited by A. S. Kechris, D. A. Martin and Y. N. Moschovakis. V, 284 pages. 1983.

Vol. 1020: Non Commutative Harmonic Analysis and Lie Groups. Proceedings, 1982. Edited by J. Carmona and M. Vergne. V, 187 pages. 1983.

Vol. 1021: Probability Theory and Mathematical Statistics. Proceedings, 1982. Edited by K. Itô and J.V. Prokhorov. VIII, 747 pages. 1983.

Vol. 1022: G. Gentili, S. Salamon and J.-P. Vigué. Geometry Seminar "Luigi Bianchi", 1982. Edited by E. Vesentini. VI, 177 pages. 1983.

Vol. 1023: S. McAdam, Asymptotic Prime Divisors. IX, 118 pages. 1983.

Vol. 1024: Lie Group Representations I. Proceedings, 1982–1983. Edited by R. Herb, R. Lipsman and J. Rosenberg. IX, 369 pages. 1983.

Vol. 1025: D. Tanré, Homotopie Rationnelle: Modèles de Chen, Quillen, Sullivan. X, 211 pages. 1983.

Vol. 1026: W. Plesken, Group Rings of Finite Groups Over p-adic Integers. V, 151 pages. 1983.

Vol. 1027: M. Hasumi, Hardy Classes on Infinitely Connected Riemann Surfaces. XII, 280 pages. 1983.

Vol. 1028: Séminaire d'Analyse P. Lelong – P. Dolbeault – H. Skoda. Années 1981/1983. Edité par P. Lelong, P. Dolbeault et H. Skoda. VIII, 328 pages. 1983.

Vol. 1029: Séminaire d'Algèbre Paul Dubreil et Marie-Paule Malliavin. Proceedings, 1982. Edité par M.-P. Malliavin. V, 339 pages. 1983.

Vol. 1030: U. Christian, Selberg's Zeta-, L-, and Eisensteinseries. XII, 196 pages. 1983.

Vol. 1031: Dynamics and Processes. Proceedings, 1981. Edited by Ph. Blanchard and L. Streit. IX, 213 pages. 1983.

Vol. 1032: Ordinary Differential Equations and Operators. Proceedings, 1982. Edited by W. N. Everitt and R. T. Lewis. XV, 521 pages. 1983.

Vol. 1033: Measure Theory and its Applications. Proceedings, 1982. Edited by J. M. Belley, J. Dubois and P. Morales. XV, 317 pages. 1983.

Vol. 1034: J. Musielak, Orlicz Spaces and Modular Spaces. V, 222 pages. 1983.

Vol. 1035: The Mathematics and Physics of Disordered Media. Proceedings, 1983. Edited by B. D. Hughes and B. W. Ninham. VII, 432 pages. 1983.

Vol. 1036: Combinatorial Mathematics X. Proceedings, 1982. Edited by L. R. A. Casse. XI, 419 pages. 1983.

Vol. 1037: Non-linear Partial Differential Operators and Quantization Procedures. Proceedings, 1981. Edited by S. I. Andersson and H.-D. Doebner. VII, 334 pages. 1983.

Vol. 1038: F. Borceux, G. Van den Bossche, Algebra in a Localic Topos with Applications to Ring Theory. IX, 240 pages. 1983.

Vol. 1039: Analytic Functions, Błażejewko 1982. Proceedings. Edited by J. Ławrynowicz. X, 494 pages. 1983

Vol. 1040: A. Good, Local Analysis of Selberg's Trace Formula. III, 128 pages. 1983.

Vol. 1041: Lie Group Representations II. Proceedings 1982–1983. Edited by R. Herb, S. Kudla, R. Lipsman and J. Rosenberg. IX, 340 pages. 1984.

Vol. 1042: A. Gut, K. D. Schmidt, Amarts and Set Function Processes. III, 258 pages. 1983.

Vol. 1043: Linear and Complex Analysis Problem Book. Edited by V. P. Havin, S. V. Hruščëv and N. K. Nikol'skii. XVIII, 721 pages. 1984.

Vol. 1044: E. Gekeler, Discretization Methods for Stable Initial Value Problems. VIII, 201 pages. 1984.

Vol. 1045: Differential Geometry. Proceedings, 1982. Edited by A. M. Naveira. VIII, 194 pages. 1984.

Vol. 1046: Algebraic K-Theory, Number Theory, Geometry and Analysis. Proceedings, 1982. Edited by A. Bak. IX, 464 pages. 1984.

Vol. 1047: Fluid Dynamics. Seminar, 1982. Edited by H. Beirão da Veiga. VII, 193 pages. 1984.

Vol. 1048: Kinetic Theories and the Boltzmann Equation. Seminar, 1981. Edited by C. Cercignani. VII, 248 pages. 1984.

Vol. 1049: B. Iochum, Cônes autopolaires et algèbres de Jordan. VI, 247 pages. 1984.

Vol. 1050: A. Prestel, P. Roquette, Formally p-adic Fields. V, 167 pages. 1984.

Vol. 1051: Algebraic Topology, Aarhus 1982. Proceedings. Edited by I. Madsen and B. Oliver. X, 665 pages. 1984.

Vol. 1052: Number Theory. Seminar, 1982. Edited by D. V. Chudnovsky, G. V. Chudnovsky, H. Cohn and M. B. Nathanson. V, 309 pages. 1984.

Vol. 1053: P. Hilton, Nilpotente Gruppen und nilpotente Räume. V, 221 pages. 1984.

Vol. 1054: V. Thomée, Galerkin Finite Element Methods for Parabolic Problems. VII, 237 pages. 1984.

Vol. 1055: Quantum Probability and Applications to the Quantum Theory of Irreversible Processes. Proceedings, 1982. Edited by L. Accardi, A. Frigerio and V. Gorini. VI, 411 pages. 1984.

Vol. 1056: Algebraic Geometry. Bucharest 1982. Proceedings, 1982. Edited by L. Bădescu and D. Popescu. VII, 380 pages. 1984.

Vol. 1057: Bifurcation Theory and Applications. Seminar, 1983. Edited by L. Salvadori. VII, 233 pages. 1984.

Vol. 1058: B. Aulbach, Continuous and Discrete Dynamics near Manifolds of Equilibria. IX, 142 pages. 1984.

Vol. 1059: Séminaire de Probabilités XVIII, 1982/83. Proceedings. Edité par J. Azéma et M. Yor. IV, 518 pages. 1984.

Vol. 1060: Topology. Proceedings, 1982. Edited by L. D. Faddeev and A. A. Mal'cev. VI, 389 pages. 1984.

Vol. 1061: Séminaire de Théorie du Potentiel. Paris, No. 7. Proceedings. Directeurs: M. Brelot, G. Choquet et J. Deny. Rédacteurs: F. Hirsch et G. Mokobodzki. IV, 281 pages. 1984.

GPSR Compliance

The European Union's (EU) General Product Safety Regulation (GPSR) is a set of rules that requires consumer products to be safe and our obligations to ensure this.

If you have any concerns about our products, you can contact us on ProductSafety@springernature.com

In case Publisher is established outside the EU, the EU authorized representative is:

Springer Nature Customer Service Center GmbH
Europaplatz 3
69115 Heidelberg, Germany

Batch number: 10704713

Printed by Printforce, the Netherlands